Developments and Applications of Block Toeplitz Iterative Solvers

T0224467

Combinatorics and Computer Science

Volume 2

Developments and Applications of Block Toeplitz Iterative Solvers

By

Xiao-Qing JIN

Faculty of Science and Technology,
University of Macau,
Macau, China.

Science Press
Beijing/New York,

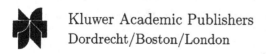
Kluwer Academic Publishers
Dordrecht/Boston/London

A C.I.P Catalogue record for this book is available from the Library of Congress.

ISBN 978-90-481-6106-5

Published by Kluwer Academic Publishers,
P. O. Box 17, 3300 AA Dordrecht, The Netherlands.

Sold and distributed in North, Central and South America
by Kluwer Academic Publishers,
101 Philip Drive, Norwell, MA 02061, U.S.A.

Sold and distributed in the People's Republic of China
by Science Press, Beijing.

In all other countries, sold and distributed
by Kluwer Academic Publishers,
P. O. Box 322, 3300 AH Dordrecht, The Netherlands.

Printed on acid-free paper

To My Family

CONTENTS

Preface

In this book we introduce current developments and applications in using iterative methods for solving block Toeplitz systems. The block Toeplitz systems arise in a variety of applications in mathematics, scientific computing and engineering, for instance, image restoration problems in image processing; numerical differential equations and integral equations; time series analysis and control theory. Krylov subspace methods and multigrid methods are proposed. One of the main results of these iterative methods is that the operation cost of solving a large class of $mn \times mn$ block Toeplitz systems is only required $O(mn \log mn)$ operations.

This book consists of twelve chapters. Various bibliographies are placed at the end of the book. In Chapter 1, we survey some background knowledge of matrix analysis and point Toeplitz iterative solvers that will be used later to develop our block Toeplitz iterative solvers.

In Chapter 2 we study block circulant preconditioners for the solution of block system $T_{mn}u = b$ by the preconditioned conjugate gradient (PCG) method where T_{mn} is an $m \times m$ block Toeplitz matrix with $n \times n$ Toeplitz blocks. The preconditioners $c_F^{(1)}(T_{mn})$, $\tilde{c}_F^{(1)}(T_{mn})$ and $c_{F,F}^{(2)}(T_{mn})$ are the matrices that preserve the block structure of T_{mn}. Specifically, they are defined to be the minimizers of $\|T_{mn} - C_{mn}\|_F$ with C_{mn} over some special classes of matrices. We prove that if T_{mn} is positive definite, then $c_F^{(1)}(T_{mn})$, $\tilde{c}_F^{(1)}(T_{mn})$ and $c_{F,F}^{(2)}(T_{mn})$ are positive definite too. We also show that they are good preconditioners for solving some special block Toeplitz systems. Finally, we briefly discuss two other preconditioners $s_{F,F}^{(2)}(T_{mn})$ and $r_{F,F}^{(2)}(T_{mn})$. The invertibility of the preconditioners $s_{F,F}^{(2)}(T_{mn})$ and $r_{F,F}^{(2)}(T_{mn})$ is studied.

In Chapter 3 block circulant preconditioners for block Toeplitz systems are studied from the viewpoint of kernels. We show that most of the well known block circulant preconditioners can be derived from convoluting the generating functions of systems with some famous kernels. The convergence analysis is also given.

In Chapter 4 we study the solutions of a block Toeplitz systems $Tu = b$ by the PCG method where $T = T_{(1)} \otimes T_{(2)} \otimes \cdots \otimes T_{(m)}$ with Toeplitz blocks

$T_{(i)} \in \mathbb{R}^{n \times n}$, $i = 1, 2, \cdots, m$. Two preconditioners C and P are proposed. The preconditioner C is a matrix that preserves the tensor structure of T and is close to T in Frobenius norm over a special class of matrices. The preconditioner P is defined for ill conditioned problems. With a fast algorithm, we show that both C and P are good preconditioners for solving block Toeplitz systems with tensor structure. Only $O(mn^m \log n)$ operations are required for the solutions of preconditioned systems. The inverse heat problem is also discussed.

In Chapter 5 we study the constrained and weighted least squares problem

$$\min_x \frac{1}{2}(b - Tx)^T W(b - Tx) \qquad .$$

where $W = \text{diag}(\omega_1, \cdots, \omega_m)$ with $\omega_1 \geq \cdots \geq \omega_m \geq 0$ and $T^T = \left(T_{(1)}^T, \cdots, T_{(k)}^T\right)$ with Toeplitz blocks $T_{(l)} \in \mathbb{R}^{n \times n}$, $l = 1, \cdots, k$. It is well known that this problem can be solved by solving the following linear system

$$\begin{cases} M\lambda + Tx = b, \\ \\ T^T \lambda = 0, \end{cases}$$

where $M = W^{-1}$. We use the PCG method with circulant-like preconditioner for solving the system and we obtain a fast convergence rate.

In Chapter 6 we study the solutions of ill conditioned block Toeplitz systems $T_{mn}u = b$ where T_{mn} are generated by a function $f(x, y) \geq 0$. Two important theorems [77], which give the relations between the values of $f(x, y)$ and the eigenvalues of T_{mn}, are proposed. Usually, the convergence rate of the conjugate gradient method for solving ill conditioned block Toeplitz systems is slow. To deal with such kind of problem, a block $\{\omega\}$-circulant preconditioner is proposed. We show that the block $\{\omega\}$-circulant preconditioner can work efficiently for ill conditioned block Toeplitz systems. A numerical comparison between the block $\{\omega\}$-circulant preconditioner and the preconditioner $c_{F,F}^{(2)}(T_{mn})$ is also given.

In Chapter 7 we first study block band Toeplitz preconditioners for the solutions of ill conditioned block Toeplitz systems $T_{mn}u = b$ by the PCG method. Here T_{mn} are assumed to be generated by a function $f(x, y) \geq 0$.

The generating function $g(x, y)$ of the block band Toeplitz preconditioners is a trigonometric polynomial of fixed degree and is determined by minimizing $|||(f - g)/f|||_\infty$. Remez algorithm is proposed to construct the preconditioners. We prove that the condition number of the preconditioned system is $O(1)$. A priori bound on the number of iterations for convergence is obtained. Finally, we briefly discuss the preconditioners based on some well known fast transforms.

In Chapter 8 we study the solutions of ill conditioned block Toeplitz systems $T_{mn}u = b$ by multigrid methods (MGMs). For a class of block Toeplitz matrices, we show that the convergence factor of the two-grid method is uniformly bounded below 1 and independent of m and n, and the full MGM has a convergence factor depending only on the number of levels. The cost per iteration for the MGM is of $O(mn \log mn)$ operations. Numerical results are given to explain the convergence rate.

In Chapter 9 we first review some results related to numerical solutions of elliptic boundary value problems. We then consider linear systems arising from implicit time discretizations and finite difference space discretizations of second-order hyperbolic equations in 2-dimensional space. We propose and analyse the use of block circulant preconditioners for the solutions of linear systems by the PCG method. For second-order hyperbolic equations with given initial and Dirichlet boundary conditions, we prove that the condition number of the preconditioned system is of $O(\alpha)$ or $O(m)$, where α is the grid ratio between the time and space steps and m is the number of interior grid points in each direction. The results are extended to parabolic equations. Numerical experiments also indicate that the preconditioned systems exhibit favorable clustering of eigenvalues that leads to a fast convergence rate. Block preconditioners based on the fast sine transform are discussed for discretized systems of second-order partial differential equations in 3-dimensional space.

In Chapter 10 block preconditioners based on the fast sine transform are proposed for solving non-symmetric and non-diagonally dominant linear systems that arise from discretizations of first-order partial differential equations. We prove that if the generalized minimal residual (GMRES) method is applied to solving the preconditioned systems, the asymptotic convergence factor of the method is independent of the mesh size and depends only on the grid ratio between the time and space steps. We compare the convergence rate of our preconditioned system with the one that preconditioned by the semi-Toeplitz preconditioner. We show that our preconditioned systems have a smaller asymptotic convergence factor and numerical experiments indicate that our preconditioned systems have a much faster convergence rate.

In Chapter 11 the block circulant preconditioner $\tilde{s}_F^{(1)}(M)$ is proposed for solving linear systems arising from numerical methods for ordinary differential equations (ODEs). We use linear multistep methods to discretize ODEs. These implicit numerical methods for solving ODEs require the solutions of non-symmetric, large and sparse linear systems at each integration step. Hence, the GMRES method is used. We show that when some stable boundary value methods are used to discretize ODEs, the preconditioner $\tilde{s}_F^{(1)}(M)$ is invertible and the eigenvalues of the preconditioned system are clustered around 1. When the GMRES method is applied to solving these preconditioned systems, we have a fast convergence rate. Numerical results are given to illustrate the effectiveness of the method. An algorithm for solving differential algebraic equations is also given.

In Chapter 12 we briefly study image restoration problems in image processing. The image of an object can be modeled as

$$g(\xi, \delta) = \int_{-\infty}^{\infty} \int_{-\infty}^{\infty} t(\xi, \delta; \alpha, \beta) f(\alpha, \beta) d\alpha d\beta + \eta(\xi, \delta)$$

where $g(\xi, \delta)$ is the degraded image, $f(\alpha, \beta)$ is the orginal image, $\eta(\xi, \delta)$ represents an additive noise. The image restoration problem is that given the observed image g, compute an approximation to the original image f. The regularized PCG least squares method with the preconditioners based on some fast transforms is proposed for solving linear systems arising from image restoration problems.

This book contains main parts of my research work in the past twelve years. Some research results are joint work with Professor Raymond H.F. Chan of the Department of Mathematics, Chinese University of Hong Kong; Professor Q.S. Chang of the Institute of Applied Mathematics, Chinese Academy of Sciences; Dr. Michael K.P. Ng of the Department of Mathematics, University of Hong Kong; Dr. H.W. Sun of the Department of Mathematics and Physics, Guangdong University of Technology; and my students Miss K.I. Kou and Mr. S.L. Lei of the Faculty of Science and Technology, University of Macau. I wish to express my sincere gratitude to my former Ph.D supervisor, Professor Raymond H.F. Chan, for leading me to this interesting area of fast iterative Toeplitz solvers and for his continual guidance, constant encouragement, long standing friendship, financial support and help. I am indebted to Professor Tony F. Chan of the Department of Mathematics, University of California, Los Angeles, for his enlightening suggestions and comments, from which I benefited a great deal during my Ph.D studies. I would like to thank my friends Professor Z.C. Shi, Dr. C.M. Cheng, Dr. C.K. Wong and Dr. M.C. Yeung for their many helpful discussions and suggestions. Thanks are also due to my parents for their encouraging and financial support. Finally, I would like to express my appreciation to my dear wife, Kathy, who eased many burdens and provided an environment and the encouragement essential to the completion of this book.

The publication of the book is supported in part by the research grants No. RG009/98-99S/JXQ/FST, No. RG010/99-00S/JXQ/FST and No. RG026/00-01S/JXQ/FST from University of Macau.

Chapter 1

Introduction

In this chapter we first introduce some background knowledge of matrix analysis which will be used throughout the book. We then give a brief survey of current developments in using preconditioned conjugate gradient (PCG) methods for solving Toeplitz systems in the point case.

1.1 Background

In this section an overview of the relevant concepts in matrix analysis is given. The material contained here will be helpful in developing our theory in later chapters.

1.1.1 Symmetric matrix, norms and tensor

A matrix $A \in \mathbb{R}^{n \times n}$ is said to be a symmetric matrix if $A^T = A$ where 'T' denotes the transposition. Real symmetric matrices have many elegant and important properties, see [95, 100], and here we present only several classical results that will be used later.

Theorem 1.1 (Spectral Theorem) *Let $A \in \mathbb{R}^{n \times n}$ be given. Then A is symmetric if and only if there exist an orthogonal matrix $Q \in \mathbb{R}^{n \times n}$ and a diagonal matrix $\Lambda \in \mathbb{R}^{n \times n}$ such that $A = Q\Lambda Q^T$.*

We recall that a matrix $M \in \mathbb{R}^{n \times n}$ is called orthogonal if $M^{-1} = M^T$.

Theorem 1.2 (Cauchy's Interlace Theorem) *Let $A \in \mathbb{R}^{n \times n}$ be a symmetric matrix with eigenvalues*

$$\lambda_1 \leq \lambda_2 \leq \cdots \leq \lambda_n$$

and let

$$\mu_1 \leq \mu_2 \leq \cdots \leq \mu_{n-1}$$

be the eigenvalues of a principal submatrix of A of order $n-1$. Then

$$\lambda_1 \leq \mu_1 \leq \lambda_2 \leq \mu_2 \leq \cdots \leq \mu_{n-1} \leq \lambda_n.$$

Theorem 1.3 (Weyl's Theorem) *Let $A, E \in \mathbb{R}^{n \times n}$ be symmetric matrices and let the eigenvalues $\lambda_i(A)$, $\lambda_i(E)$ and $\lambda_i(A+E)$ be arranged in increasing order. Then for each $k = 1, 2, \cdots, n$, we have*

$$\lambda_k(A) + \lambda_1(E) \leq \lambda_k(A+E) \leq \lambda_k(A) + \lambda_n(E).$$

Theorem 1.4 (Courant–Fischer's Minimax Theorem) *Let $A \in \mathbb{R}^{n \times n}$ be a symmetric matrix with eigenvalues*

$$\lambda_1 \leq \lambda_2 \leq \cdots \leq \lambda_n$$

and let k be a given integer with $1 \leq k \leq n$. Then

$$\lambda_k = \min_{\dim \mathcal{X}=k} \max_{0 \neq x \in \mathcal{X}} \frac{x^T A x}{x^T x} = \min_{\dim \mathcal{X}=n-k+1} \max_{0 \neq x \in \mathcal{X}} \frac{x^T A x}{x^T x}.$$

In particular, for the smallest and largest eigenvalues, we have

$$\lambda_1 = \min_{x \neq 0} \frac{x^T A x}{x^T x} \quad and \quad \lambda_n = \max_{x \neq 0} \frac{x^T A x}{x^T x}.$$

The results of Spectral Theorem, Cauchy's Interlace Theorem, Weyl's Theorem and Courant–Fischer's Minimax Theorem can be extended to the case of Hermitian matrices. We remark that a matrix $A \in \mathbb{C}^{n \times n}$ is said to be a Hermitian matrix if $A^* = A$ where '$*$' denotes the conjugate transposition. For any arbitrary $A \in \mathbb{C}^{n \times n}$, it is possible to decompose A into an 'almost diagonal form' – the Jordan canonical form.

Theorem 1.5 (Jordan's Decomposition Theorem) *If $A \in \mathbb{C}^{n \times n}$, then there exists an invertible matrix $X \in \mathbb{C}^{n \times n}$ such that*

$$X^{-1}AX = J \equiv \mathrm{diag}(J_1, J_2, \cdots, J_k)$$

which is called the Jordan canonical form of A, where $\mathrm{diag}(\cdot)$ denotes the diagonal matrix and

$$J_i = \begin{pmatrix} \lambda_i & 1 & 0 & \cdots & 0 \\ 0 & \lambda_i & 1 & \ddots & \vdots \\ \vdots & 0 & \ddots & \ddots & 0 \\ \vdots & & \ddots & \ddots & 1 \\ 0 & \cdots & \cdots & 0 & \lambda_i \end{pmatrix} \in \mathbb{C}^{n_i \times n_i},$$

for $i = 1, 2, \cdots, k$, are called Jordan blocks with $n_1 + \cdots + n_k = n$. The Jordan canonical form of A is unique up to the permutation of diagonal Jordan blocks. The eigenvalues $\lambda_i, i = 1, 2, \cdots, k$, are not necessarily distinct. If $A \in \mathbb{R}^{n \times n}$ with only real eigenvalues, then the matrix X can be taken to be real.

Let

$$x = (x_1, x_2, \cdots x_n)^T \in \mathbb{C}^n.$$

A vector norm on $\mathbb{C}^n$ is a function that assign to each $x \in \mathbb{C}^n$ a real number $\|x\|$, called the norm of x, such that the following three properties are satisfied for all $x, y \in \mathbb{C}^n$ and all $\alpha \in \mathbb{C}$:

(i) $\|x\| > 0$ if $x \neq 0$ and $\|0\| = 0$;
(ii) $\|\alpha x\| = |\alpha| \|x\|$;
(iii) $\|x + y\| \leq \|x\| + \|y\|$.

A useful class of vector norms is the p-norm defined by

$$\|x\|_p \equiv \left(\sum_{i=1}^n |x_i|^p \right)^{\frac{1}{p}}.$$

The following p-norms are the most commonly used norms in practice:

$$\|x\|_1 = \sum_{i=1}^n |x_i|, \quad \|x\|_2 = \left(\sum_{i=1}^n |x_i|^2 \right)^{1/2}, \quad \|x\|_\infty = \max_{1 \leq i \leq n} |x_i|.$$

Cauchy–Schwarz's inequality concerning $\|\cdot\|_2$ is given as follows,

$$|x^*y| \leq \|x\|_2 \|y\|_2$$

for $x, y \in \mathbb{C}^n$. A very important property of vector norms on $\mathbb{C}^n$ is that all vector norms on $\mathbb{C}^n$ are equivalent, i.e., if $\|\cdot\|_\alpha$ and $\|\cdot\|_\beta$ are two norms on $\mathbb{C}^n$, then there exist two positive constants c_1 and c_2 such that

$$c_1 \|x\|_\alpha \leq \|x\|_\beta \leq c_2 \|x\|_\alpha$$

for all $x \in \mathbb{C}^n$. For example, if $x \in \mathbb{C}^n$, then we have

$$\|x\|_2 \leq \|x\|_1 \leq \sqrt{n} \|x\|_2,$$

$$\|x\|_\infty \leq \|x\|_2 \leq \sqrt{n} \|x\|_\infty$$

and

$$\|x\|_\infty \leq \|x\|_1 \leq n \|x\|_\infty.$$

Let

$$A = (a_{i,j})_{i,j=1}^n \in \mathbb{C}^{n \times n}.$$

We now turn our attention to matrix norms. A matrix norm is a function that assigns to each $A \in \mathbb{C}^{n \times n}$ a real number $\|A\|$, called the norm of A, such that the following four properties are satisfied for all $A, B \in \mathbb{C}^{n \times n}$ and all $\alpha \in \mathbb{C}$:

(i) $\|A\| > 0$ if $A \neq 0$ and $\|0\| = 0$;
(ii) $\|\alpha A\| = |\alpha| \|A\|$;
(iii) $\|A + B\| \leq \|A\| + \|B\|$;
(iv) $\|AB\| \leq \|A\| \|B\|$.

For every vector norm, we can define a matrix norm in a natural way. Given the vector norm $\|\cdot\|_v$, the matrix norm induced by $\|\cdot\|_v$ is defined by

$$\|A\|_v \equiv \max_{x \neq 0} \frac{\|Ax\|_v}{\|x\|_v}.$$

The most important matrix norms are the matrix p-norms induced by the vector p-norms for $p = 1, 2, \infty$:

$$\|A\|_1 = \max_{1 \leq j \leq n} \sum_{i=1}^n |a_{i,j}|, \quad \|A\|_2 = \sigma_{\max}(A), \quad \|A\|_\infty = \max_{1 \leq i \leq n} \sum_{j=1}^n |a_{i,j}|,$$

where $\sigma_{\max}(A)$ denotes the largest singular value of A, see [38]. The Frobenius norm is defined by

$$\|A\|_F \equiv \left(\sum_{j=1}^{n} \sum_{i=1}^{n} |a_{i,j}|^2 \right)^{1/2}.$$

One of the most important properties of $\| \cdot \|_2$ and $\| \cdot \|_F$ is that for any unitary matrices Q and Z,

$$\|A\|_2 = \|QAZ\|_2$$

and

$$\|A\|_F = \|QAZ\|_F.$$

We recall that a matrix $M \in \mathbb{C}^{n \times n}$ is called unitary if $M^{-1} = M^*$.

Let $A = (a_{i,j}) \in \mathbb{C}^{m \times n}$ and $B \in \mathbb{C}^{p \times q}$. The $mp \times nq$ matrix

$$A \otimes B \equiv \begin{pmatrix} a_{11}B & a_{12}B & \cdots & a_{1n}B \\ a_{21}B & a_{22}B & \cdots & a_{2n}B \\ \vdots & \vdots & & \vdots \\ a_{m1}B & a_{m2}B & \cdots & a_{mn}B \end{pmatrix}$$

is called the tensor product of A and B.

The basic properties of the tensor product are summarized in the following theorem.

Theorem 1.6 *Let* $A \in \mathbb{C}^{m \times n}$ *and* $B \in \mathbb{C}^{p \times q}$. *Then we have,*

(i) $rank(A \otimes B) = rank(A) \cdot rank(B)$;

(ii) $(A \otimes B)^* = A^* \otimes B^*$;

(iii) $(A \otimes B)(C \otimes D) = (AC) \otimes (BD)$, *where* $C \in \mathbb{C}^{n \times k}$, $D \in \mathbb{C}^{q \times r}$;

(iv) *If both* A *and* B *are invertible, then* $A \otimes B$ *is also invertible and*

$$(A \otimes B)^{-1} = A^{-1} \otimes B^{-1}.$$

1.1.2   Condition number and error estimates

When we solve a linear system $Ax = b$, a good measurement, which can tell us how sensitive the computed solution is to the input perturbations, is needed. The condition number of matrix is then defined. It relates the perturbations in x to the perturbations in A and b.

Definition 1.1 *Let $\|\cdot\|$ be any p-norm of matrix and A be an invertible matrix. The condition number of A is defined as follows,*

$$\kappa(A) \equiv \|A\|\|A^{-1}\|. \tag{1.1}$$

Obviously, the condition number depends on the matrix norm used. Since

$$1 = \|I\| = \|A \cdot A^{-1}\| \leq \|A\| \cdot \|A^{-1}\|$$

where I is the identity matrix, it follows that $\kappa(A) \geq 1$. When $\kappa(A)$ is small, then A is said to be well conditioned, whereas if $\kappa(A)$ is large, then A is said to be ill conditioned.

Let $\hat{x}$ be an approximation of the exact solution x of $Ax = b$. The error vector is defined as follows,

$$e = x - \hat{x},$$

i.e.,

$$x = \hat{x} + e. \tag{1.2}$$

The absolute error is given by

$$\|e\| = \|x - \hat{x}\|$$

for any vector norm. If $x \neq 0$, then the relative error is defined by

$$\frac{\|e\|}{\|x\|} = \frac{\|x - \hat{x}\|}{\|x\|}.$$

We have by substituting (1.2) into $Ax = b$,

$$A(\hat{x} + e) = A\hat{x} + Ae = b.$$

Therefore,

$$A\hat{x} = b - Ae = \hat{b}.$$

The $\hat{x}$ is the exact solution of $A\hat{x} = \hat{b}$ where $\hat{b}$ is the perturbation of vector b. Since $x = A^{-1}b$ and $\hat{x} = A^{-1}\hat{b}$, we have

$$\|x - \hat{x}\| = \|A^{-1}(b - \hat{b})\| \leq \|A^{-1}\| \cdot \|b - \hat{b}\|. \tag{1.3}$$

Similarly,

$$\|b\| = \|Ax\| \leq \|A\| \cdot \|x\|,$$

i.e.,

$$\frac{1}{\|x\|} \leq \frac{\|A\|}{\|b\|}. \tag{1.4}$$

Combining (1.3), (1.4) and (1.1), we obtain the following theorem which gives the effect of perturbations of the vector b on the solution of $Ax = b$ in terms of the condition number.

Theorem 1.7 *Let $\hat{x}$ be an approximate solution of $Ax = b$. Then*

$$\frac{\|x - \hat{x}\|}{\|x\|} \leq \kappa(A)\frac{\|b - \hat{b}\|}{\|b\|}.$$

The next theorem includes the effect of perturbations of the coefficient matrix A on the solution of $Ax = b$ in terms of the condition number.

Theorem 1.8 *Let A be an invertible matrix and $\hat{A}$ be the perturbed matrix of A such that*

$$\|A - \hat{A}\| \cdot \|A^{-1}\| < 1.$$

If $Ax = b$ and $\hat{A}\hat{x} = \hat{b}$, then

$$\frac{\|x - \hat{x}\|}{\|x\|} \leq \frac{\kappa(A)}{1 - \kappa(A)\frac{\|A-\hat{A}\|}{\|A\|}} \left(\frac{\|A - \hat{A}\|}{\|A\|} + \frac{\|b - \hat{b}\|}{\|b\|} \right).$$

Proof: Let

$$E = A - \hat{A} \quad \text{and} \quad \beta = b - \hat{b}.$$

By subtracting $Ax = b$ from $\hat{A}\hat{x} = \hat{b}$, we have

$$A(x - \hat{x}) = -E\hat{x} + \beta.$$

Furthermore, we get

$$\frac{\|x - \hat{x}\|}{\|x\|} \leq \|A^{-1}E\|\frac{\|\hat{x}\|}{\|x\|} + \|A^{-1}\|\frac{\|Ax\|}{\|x\|}\frac{\|\beta\|}{\|b\|}.$$

By using

$$\|\hat{x}\| \leq \|\hat{x} - x\| + \|x\| \quad \text{and} \quad \|Ax\| \leq \|A\|\|x\|,$$

we then have

$$\frac{\|x - \hat{x}\|}{\|x\|} \leq \|A^{-1}E\| \frac{\|x - \hat{x}\|}{\|x\|} + \|A^{-1}E\| + \|A^{-1}\|\|A\| \frac{\|\beta\|}{\|b\|},$$

i.e.,

$$(1 - \|A^{-1}E\|) \frac{\|x - \hat{x}\|}{\|x\|} \leq \|A^{-1}E\| + \kappa(A) \frac{\|\beta\|}{\|b\|}.$$

Since

$$\|A^{-1}E\| \leq \|A^{-1}\|\|E\| = \|A^{-1}\|\|A - \hat{A}\| < 1,$$

we get

$$\frac{\|x - \hat{x}\|}{\|x\|} \leq (1 - \|A^{-1}E\|)^{-1} \left(\|A^{-1}E\| + \kappa(A) \frac{\|\beta\|}{\|b\|} \right).$$

By using

$$\|A^{-1}E\| \leq \|A^{-1}\|\|E\| = \kappa(A) \frac{\|E\|}{\|A\|},$$

we finally have

$$\frac{\|x - \hat{x}\|}{\|x\|} \leq \frac{\kappa(A)}{1 - \kappa(A)\frac{\|E\|}{\|A\|}} \left(\frac{\|E\|}{\|A\|} + \frac{\|\beta\|}{\|b\|} \right).$$

The proof is complete.   □

Theorems 1.7 and 1.8 give the upper bounds for the relative error of x in terms of the condition number of A. From the Theorems 1.7 and 1.8, we know that if A is well conditioned, i.e., $\kappa(A)$ is small, the relative error in x will be small if the relative errors in both A and b are small.

1.1.3   Toeplitz matrix and circulant matrix

An $n \times n$ Toeplitz matrix is of the following form

$$T_n = \begin{pmatrix} t_0 & t_{-1} & \cdots & t_{2-n} & t_{1-n} \\ t_1 & t_0 & t_{-1} & \cdots & t_{2-n} \\ \vdots & t_1 & t_0 & \ddots & \vdots \\ t_{n-2} & \cdots & \ddots & \ddots & t_{-1} \\ t_{n-1} & t_{n-2} & \cdots & t_1 & t_0 \end{pmatrix}, \tag{1.5}$$

i.e., T_n is constant along its diagonals. The name of Toeplitz is in memory of Toeplitz's early work in 1911, see [10] and [23]. Since there is a variety of applications in mathematics, scientific computing and engineering, these applications have motivated mathematicians, scientists and engineers to develop specifically fast algorithms for solving Toeplitz systems $T_n u = b$. These algorithms are called Toeplitz solvers.

In 1986, Strang [87] and Olkin [70] proposed independently the use of the PCG method with circulant matrices as preconditioners to solve Toeplitz systems. The circulant matrix is defined as follows:

$$
C_n = \begin{pmatrix}
c_0 & c_{-1} & \cdots & c_{2-n} & c_{1-n} \\
c_1 & c_0 & c_{-1} & \cdots & c_{2-n} \\
\vdots & c_1 & c_0 & \ddots & \vdots \\
c_{n-2} & \cdots & \ddots & \ddots & c_{-1} \\
c_{n-1} & c_{n-2} & \cdots & c_1 & c_0
\end{pmatrix}
$$

where $c_{-k} = c_{n-k}$ for $1 \leq k \leq n - 1$. It is well known that circulant matrices can be diagonalized by the Fourier matrix F_n, see [33], i.e.,

$$
C_n = F_n^* \Lambda_n F_n, \tag{1.6}
$$

where the entries of F_n are given by

$$
(F_n)_{j,k} = \frac{1}{\sqrt{n}} e^{2\pi i jk/n}, \qquad i \equiv \sqrt{-1},
$$

for $0 \leq j, k \leq n - 1$, and Λ_n is a diagonal matrix holding the eigenvalues of C_n. For the Fourier matrix F_n, when there is no ambiguity, we shall denote F.

1.1.4 Conjugate gradient method

The scheme of the conjugate gradient method, one of the most popular and successful iterative methods for solving Hermitian positive definite systems $H_n x = b$, is given as follows, see [38, 75]. At the initialization step, we choose $x^{(0)}$, calculate

$$
g^{(0)} = H_n x^{(0)} - b,
$$

and put $d^{(0)} = -g^{(0)}$. In the iteration steps, we have

$$
\begin{cases}
\tau_k := \dfrac{g^{(k)^*} g^{(k)}}{d^{(k)^*} H_n d^{(k)}}, \\[3ex]
x^{(k+1)} := x^{(k)} + \tau_k d^{(k)}, \\[3ex]
g^{(k+1)} := g^{(k)} + \tau_k H_n d^{(k)}, \\[3ex]
\beta_k := \dfrac{g^{(k+1)^*} g^{(k+1)}}{g^{(k)^*} g^{(k)}}, \\[3ex]
d^{(k+1)} := -g^{(k+1)} + \beta_k d^{(k)},
\end{cases}
$$

where $d^{(k)}$, $g^{(k)}$ are vectors and τ_k, β_k are scalars, $k = 0, 1, \cdots$. The $x^{(k)}$ is the approximation to the true solution after the k-th iteration.

For the convergence rate of the conjugate gradient method, we have the following theorem, see [2] and [99].

Theorem 1.9 *Let $x^{(k)}$ be the k-th iterant of the conjugate gradient method applied to the system $H_n x = b$ and let x be the true solution of the system. If the eigenvalues λ_j of H_n are ordered such that*

$$0 < \lambda_1 \le \dots \le \lambda_p \le b_1 \le \lambda_{p+1} \le \dots \le \lambda_{n-q} \le b_2 \le \lambda_{n-q+1} \le \dots \le \lambda_n$$

where b_1 and b_2 are two constants, then

$$\frac{\|x - x^{(k)}\|}{\|x - x^{(0)}\|} \le 2 \left(\frac{\alpha - 1}{\alpha + 1} \right)^{k-p-q} \cdot \max_{\lambda \in [b_1, b_2]} \prod_{j=1}^{p} \left(\frac{\lambda - \lambda_j}{\lambda_j} \right) .$$

Here $\|\cdot\|$ is the energy norm given by $\|v\|^2 \equiv v^* H_n v$ and $\alpha \equiv (b_2/b_1)^{1/2} \ge 1$.

From Theorem 1.9 we note that when n is increased, if p, q are constants which not depend on n, and if λ_1 is uniformly bounded from zero, then the convergence rate is linear, i.e., the number of iterations is independent of n. We also notice that the more clustered the eigenvalues are, the faster the convergence rate will be.

Corollary 1.1 *Let $x^{(k)}$ be the k-th iterant of the conjugate gradient method applied to the system $H_n x = b$ and let x be the true solution of the system. If the eigenvalues λ_j of H_n are ordered such that*

$$0 < \delta \leq \lambda_1 \leq \dots \leq \lambda_p \leq 1 - \epsilon \leq \lambda_{p+1} \leq \dots$$

$$\leq \lambda_{n-q} \leq 1 + \epsilon \leq \lambda_{n-q+1} \leq \dots \leq \lambda_n$$

where $0 < \epsilon < 1$, then

$$\frac{\|x - x^{(k)}\|}{\|x - x^{(0)}\|} \leq 2 \left(\frac{1 + \epsilon}{\delta} \right)^p \epsilon^{k-p-q}$$

where $\| \cdot \|$ is the energy norm and $k \geq p + q$.

Proof: For α given in Theorem 1.9, we have

$$\alpha \equiv \left(\frac{b_2}{b_1} \right)^{\frac{1}{2}} = \left(\frac{1 - \epsilon}{1 + \epsilon} \right)^{\frac{1}{2}}.$$

Therefore,

$$\frac{\alpha - 1}{\alpha + 1} = \frac{1 - \sqrt{1 - \epsilon^2}}{\epsilon} < \epsilon.$$

For $1 \leq j \leq p$ and $\lambda \in [1 - \epsilon, 1 + \epsilon]$, we have

$$0 \leq \frac{\lambda - \lambda_j}{\lambda_j} \leq \frac{1 + \epsilon}{\delta}.$$

Thus, by using Theorem 1.9, we obtain

$$\frac{\|x - x^{(k)}\|}{\|x - x^{(0)}\|} \leq 2 \left(\frac{\alpha - 1}{\alpha + 1} \right)^{k-p-q} \cdot \max_{\lambda \in [b_1, b_2]} \prod_{j=1}^{p} \left(\frac{\lambda - \lambda_j}{\lambda_j} \right)$$

$$\leq 2 \left(\frac{1 + \epsilon}{\delta} \right)^p \epsilon^{k-p-q}. \quad \square$$

In order to accelerate the convergence rate, we usually precondition the system, i.e., instead of solving the original system, we solve the following preconditioned system

$$M_n^{-1} H_n x = M_n^{-1} b.$$

The preconditioner M_n is Hermitian positive definite and chosen with two criteria in mind, see [2] and [38]:

I. $M_n r = d$ is easy to solve;

II. the spectrum of $M_n^{-1} H_n$ is clustered and (or) $M_n^{-1} H_n$ is well conditioned compared to H_n.

The main work involved in implementing the conjugate gradient method to the preconditioned system $M_n^{-1} H_n x = M_n^{-1} b$ is the matrix-vector product $M_n^{-1} H_n v$ for some vector v.

In [87] and [70] Strang and Olkin noted that for any Toeplitz matrix T_n with a circulant preconditioner C_n, this product $C_n^{-1} T_n v$ can be computed in $O(n \log n)$ operations for any vector v as circulant systems can be solved efficiently by the Fast Fourier Transform (FFT) and the multiplication $T_n v$ can also be computed by FFTs by first embedding T_n into a $2n \times 2n$ circulant matrix. More precisely, we have a $2n \times 2n$ circulant matrix with T_n embedded inside as follows,

$$\begin{pmatrix} T_n & \times \\ \times & T_n \end{pmatrix} \begin{pmatrix} v \\ 0 \end{pmatrix} = \begin{pmatrix} T_n v \\ \dagger \end{pmatrix},$$

and then the multiplication can be carried out by using the decomposition as in (1.6). The operation cost is, therefore, $O(2n \log(2n))$. Thus, the cost per iteration of the PCG method is still $O(n \log n)$.

In particular, if the method converges linearly or superlinearly, then the complexity of the algorithm remains $O(n \log n)$. This is one of main important results of this algorithm as compared to the operation cost of $O(n \log^2 n)$ required by fast direct Toeplitz solvers, see [1, 62]. For the stories of direct Toeplitz solvers, we refer to [1, 9, 38, 43, 62].

1.1.5 Generating function and spectral analysis

In order to analyse the convergence rate of the PCG method, we need to introduce the following definition of clustered spectrum and the technical term of generating function, see [23].

Definition 1.2 *A sequence of matrices $\{A_n\}_{n=1}^{\infty}$ is said to have clustered spectra around 1 : if for any $\epsilon > 0$, there exist N, $M > 0$, such that for any $n > N$, at most M eigenvalues of $A_n - I_n$ have absolute value larger than ϵ. Here $I_n \in \mathbb{R}^{n \times n}$ denotes the identity matrix.*

When there is no ambiguity, we shall write I to denote the identity matrix. From Theorem 1.9 we know that the convergence rate of the PCG method with circulant preconditioner C_n for solving the Toeplitz system $T_n u = b$ depends on the spectrum of $C_n^{-1} T_n$ and is a function of n. To link all the T_n's together we assume that the diagonals $\{t_k\}_{k=-n+1}^{n-1}$ of T_n are the Fourier coefficients of a function f, i.e.,

$$t_k(f) \equiv \frac{1}{2\pi} \int_{-\pi}^{\pi} f(x) e^{-ikx} dx, \qquad i \equiv \sqrt{-1}.$$

The f is called the generating function of T_n. In practical problems from industry and engineering, we are usually given f first, not Toeplitz matrices T_n, see [23].

The function f is assumed in certain class of functions such that all the T_n are invertible. We note that:

(i) **When f is real-valued, then $T_n(f)$ are Hermitian.**

(ii) **When f is real-valued and even, then $T_n(f)$ are real symmetric.**

Let $\mathcal{C}_{2\pi}$ be the set of all 2π-periodic continuous real-valued functions defined on $[-\pi, \pi]$. The following theorem gives the relation between the values of f and the eigenvalues of $T_n(f)$, see [40].

Theorem 1.10 *Let T_n be given by (1.5) with a generating function $f(x) \in \mathcal{C}_{2\pi}$. Let $\lambda_{\min}(T_n)$ and $\lambda_{\max}(T_n)$ denote the smallest and largest eigenvalues of T_n respectively. Then we have*

$$f_{\min} \leq \lambda_{\min}(T_n) \leq \lambda_{\max}(T_n) \leq f_{\max}$$

where $f_{\min}$ and $f_{\max}$ denote the minimum and maximum values of $f(x)$ respectively. In particular, if $f_{\min} > 0$, then T_n is positive definite.

The following theorem is an improvement on Theorem 1.10. The proof of the theorem can be found in [16].

Theorem 1.11 *Let $f(x) \in C_{2\pi}$. If $f_{\min} < f_{\max}$. Then for all $n > 0$,*

$$f_{\min} < \lambda_i(T_n(f)) < f_{\max}, \quad i = 1, \cdots, n,$$

where $\lambda_i(T_n)$ is the i-th eigenvalue of T_n. In particular, if $f \geq 0$, then $T_n(f)$ are positive definite for all n. Moreover,

$$\lim_{n \to \infty} \lambda_{\max}(T_n) = f_{\max} \quad and \quad \lim_{n \to \infty} \lambda_{\min}(T_n) = f_{\min}.$$

From Theorem 1.11, we know that if $f \geq 0$, then $T_n(f)$ is always positive definite. When f vanishes at some point $x_0 \in [-\pi, \pi]$, then the condition number $\kappa(T_n)$ of T_n is unbounded as n tend to infinity, i.e., T_n is ill conditioned.

1.2 Circulant preconditioners

Since 1986, a lot of circulant preconditioners have been proposed for solving Toeplitz systems. We introduce some of them which have been proved to be good preconditioners.

1.2.1 Strang's preconditioner

For T_n given by (1.5) Strang's preconditioner $s(T_n)$ is defined to be the circulant matrix obtained by copying the central diagonals of T_n and bringing them around to complete the circulant requirement. More precisely, the diagonals of $s(T_n)$ are given by

$$s_k = \begin{cases} t_k, & 0 \leq k \leq \lfloor n/2 \rfloor, \\ t_{k-n}, & \lfloor n/2 \rfloor < k < n, \\ s_{n+k}, & 0 < -k < n. \end{cases} \tag{1.7}$$

For the convergence rate of the PCG method, we have the following theorem, see [14].

Theorem 1.12 *Let $f > 0$ be the generating function of T_n in the Wiener class, i.e.,*

$$\sum_{k=0}^{\infty} |t_k(f)| < \infty.$$

Then the spectra of $(s(T_n))^{-1}T_n$ are clustered around 1 for large n.

Using the standard error analysis of the PCG method and Corollary 1.1, one can easily prove that the convergence rate is superlinear, see [26], i.e.,

$$\lim_{q \to \infty} \frac{\|e^{(q+1)}\|}{\|e^{(q)}\|} = 0.$$

Here $e^{(q)}$ is the error vector after the q-th iteration and $\|\cdot\|$ is the energy norm.

1.2.2   T. Chan's preconditioner

Since the preconditioner proposed by T. Chan is defined not just for Toeplitz matrices but for general matrices as well, we then begin with the general case. Given a unitary matrix $U \in \mathbb{C}^{n \times n}$, let

$$\mathcal{M}_U = \{U^* \Lambda_n U \mid \Lambda_n \text{ is an } n \times n \text{ diagonal matrix}\}. \qquad (1.8)$$

We note that in (1.8), when $U = F$, the Fourier matrix, $\mathcal{M}_F$ is the set of all circulant matrices, see [33]. Let $\delta(A_n)$ denote the diagonal matrix whose diagonal is equal to the diagonal of the matrix A_n. The following lemma was first proved by R. Chan, Jin and Yeung [21] for the case $U = F$ and was extended to the general unitary case by Huckle [50].

Lemma 1.1 *For any arbitrary* $A_n = (a_{pq}) \in \mathbb{C}^{n \times n}$, *let* $c_U(A_n)$ *be the minimizer of* $\|W_n - A_n\|_F$ *over all* $W_n \in \mathcal{M}_U$. *Then*

(i) $c_U(A_n)$ *is uniquely determined by* A_n *and is given by*

$$c_U(A_n) = U^* \delta(U A_n U^*) U. \qquad (1.9)$$

(ii) *If* A_n *is Hermitian, then* $c_U(A_n)$ *is also Hermitian. Furthermore, if* $\lambda_{\min}(\cdot)$ *and* $\lambda_{\max}(\cdot)$ *denote the smallest and largest eigenvalues respectively, then we have*

$$\lambda_{\min}(A_n) \leq \lambda_{\min}(c_U(A_n)) \leq \lambda_{\max}(c_U(A_n)) \leq \lambda_{\max}(A_n).$$

In particular, if A_n *is positive definite, then so is* $c_U(A_n)$.

(iii) *The c_U is a linear projection operator from $\mathbb{C}^{n \times n}$ into $\mathcal{M}_U$ and has the operator norms*

$$\|c_U\|_2 = \sup_{\|A_n\|_2 = 1} \|c_U(A_n)\|_2 = 1$$

and

$$\|c_U\|_F = \sup_{\|A_n\|_F = 1} \|c_U(A_n)\|_F = 1.$$

(iv) *When U is the Fourier matrix F, we then have*

$$c_F(A_n) = \sum_{j=0}^{n-1} \left(\frac{1}{n} \sum_{p-q \equiv j \,(\bmod\, n)} a_{pq} \right) Q^j , \qquad (1.10)$$

where Q is an $n \times n$ circulant matrix given by

$$Q \equiv \begin{pmatrix} 0 & & & & 1 \\ 1 & 0 & & & \\ 0 & 1 & \ddots & & \\ \vdots & \ddots & \ddots & \ddots & \\ 0 & \cdots & 0 & 1 & 0 \end{pmatrix}. \qquad (1.11)$$

Proof: We only prove (i) and (ii). For (iii) and (iv), we refer to [21, 50].

(i) Since the Frobenius norm is unitary invariant, we have

$$\|W_n - A_n\|_F = \|U^* \Lambda_n U - A_n\|_F = \|\Lambda_n - U A_n U^*\|_F.$$

Thus the problem of minimizing $\|W_n - A_n\|_F$ over $\mathcal{M}_U$ is equivalent to the problem of minimizing $\|\Lambda_n - U A_n U^*\|_F$ over all diagonal matrices. Since Λ_n can only affect the diagonal entries of $U A_n U^*$, we see that the solution for the latter problem is $\Lambda_n = \delta(U A_n U^*)$. Hence $U^* \delta(U A_n U^*) U$ is the minimizer of $\|W_n - A_n\|_F$. It is clear from the argument above that Λ_n and hence $c_U(A_n)$ are uniquely determined by A_n.

(ii) It is clear that $c_U(A_n)$ is Hermitian when A_n is Hermitian. By (i), we know that the eigenvalues of $c_U(A_n)$ are given by $\delta(U A_n U^*)$. Suppose that

$$\delta(U A_n U^*) = \mathrm{diag}(\lambda_0, \cdots, \lambda_{n-1})$$

with $\lambda_j = \lambda_{\min}(c_U(A_n))$ and $\lambda_k = \lambda_{\max}(c_U(A_n))$. Let e_j, $e_k \in \mathbb{R}^n$ denote the j-th and the k-th unit vectors respectively. Since A_n is Hermitian, by Courant-Fischer's Minimax Theorem, we have

$$\lambda_{\max}(c_U(A_n)) = \lambda_k = \frac{e_k^* U A_n U^* e_k}{e_k^* e_k}$$

$$\leq \max_{x \neq 0} \frac{x^* U A_n U^* x}{x^* x} = \max_{x \neq 0} \frac{x^* A_n x}{x^* x} = \lambda_{\max}(A_n).$$

Similarly,

$$\lambda_{\min}(A_n) = \min_{x \neq 0} \frac{x^* A_n x}{x^* x} = \min_{x \neq 0} \frac{x^* U A_n U^* x}{x^* x}$$

$$\leq \frac{e_j^* U A_n U^* e_j}{e_j^* e_j} = \lambda_j = \lambda_{\min}(c_U(A_n)). \quad \square$$

The $c_F(A_n)$, called the optimal circulant preconditioner, is first proposed by T. Chan in [30]. It has been proved to be a good preconditioner for solving a large class of Toeplitz systems by the PCG method, see [21]. For T_n given by (1.5), the diagonals of $c_F(T_n)$ are given by

$$c_k = \begin{cases} \dfrac{(n-k)t_k + kt_{k-n}}{n}, & 0 \leq k < n, \\ c_{n+k}, & 0 < -k < n. \end{cases} \tag{1.12}$$

For the convergence rate of the PCG method, we have the following theorem, see [28].

Theorem 1.13 *Let $f > 0$ be the generating function of T_n in $\mathcal{C}_{2\pi}$. Then the spectra of $(c_F(T_n))^{-1} T_n$ are clustered around 1 for large n. Thus, the convergence rate of the PCG method is superlinear.*

Comparing with Theorem 1.12, we remark that the Wiener class of functions is a proper subset of $\mathcal{C}_{2\pi}$.

1.2.3   R. Chan's preconditioner

R. Chan's preconditioner $r(T_n)$ proposed in [14] is defined as follows. For T_n given by (1.5), the preconditioner $r(T_n)$ is the circulant matrix with diagonals:

$$
r_k = \begin{cases} t_{k-n} + t_k, & 0 \le k < n, \\ r_{n+k}, & 0 < -k < n, \end{cases} \tag{1.13}
$$

where t_{-n} is taken to be 0. For the convergence rate of the PCG method, we have the following theorem, see [14].

Theorem 1.14 *Let $f > 0$ be the generating function of T_n in the Wiener class. Then the spectra of $(r(T_n))^{-1}T_n$ are clustered around 1 for large n. Thus, the convergence rate of the PCG method is superlinear.*

1.2.4   Huckle's preconditioner

For T_n given by (1.5), Huckle's preconditioner H_n^p is defined to be the circulant matrix with eigenvalues

$$
\lambda_k(H_n^p) = \sum_{j=-p+1}^{p-1} t_j \left(1 - \frac{|j|}{p}\right) e^{2\pi i j k/n}, \quad k = 0, \cdots, n-1. \tag{1.14}
$$

When $p = n$, it is nothing new but T. Chan's preconditioner. For the convergence rate of the PCG method, we have the following theorem, see [50].

Theorem 1.15 *Let $f > 0$ be the generating function of T_n with Fourier coefficients $t_k(f)$ that satisfy*

$$
\sum_{k=0}^{\infty} |k| |t_k(f)|^2 < \infty.
$$

Then the spectra of $(H_n^p)^{-1}T_n$ are clustered around 1 for large n. Thus, the convergence rate of the PCG method is superlinear.

1.3   Non-circulant preconditioners

In this section, we introduce some other preconditioners which are not circulant for solving Toeplitz systems.

1.3.1 Optimal preconditioners based on fast transforms

In addition to the FFT, a lot of fast transforms are used in scientific computing and engineering. With the U in (1.8) taking other matrices based on different fast transforms, we then have new classes of optimal preconditioners for solving Toeplitz systems.

Optimal preconditioner based on the sine transform

Let $\mathcal{S}$ be the set of all $n \times n$ matrices that can be diagonalized by the discrete sine transform matrix Φ^s, i.e.,

$$\mathcal{S} = \{\Phi^s \Lambda_n \Phi^s \mid \Lambda_n \text{ is an } n \times n \text{ diagonal matrix}\}.$$

Here the (j, k)-th entry of Φ^s is given by

$$\sqrt{\frac{2}{n+1}} \sin\left(\frac{\pi j k}{n+1}\right),$$

for $1 \le j, k \le n$. Given any arbitrary matrix $A_n \in \mathbb{C}^{n \times n}$, we define an operator Ψ_s which maps A_n to $\Psi_s(A_n)$ that minimizes $\|A_n - B_n\|_F$ over all $B_n \in \mathcal{S}$, see [7]. For the construction of $\Psi_s(A_n)$, we refer to [25].

Optimal preconditioner based on cosine transform

Let $\mathcal{C}$ be the set of all $n \times n$ matrices that can be diagonalized by the discrete cosine transform matrix Φ^c, i.e.,

$$\mathcal{C} = \{(\Phi^c)^T \Lambda_n \Phi^c \mid \Lambda_n \text{ is an } n \times n \text{ diagonal matrix}\}.$$

Here the (j, k)-th entry of Φ^c is given by

$$\sqrt{\frac{2 - \delta_{j1}}{n}} \cos\left(\frac{(j-1)(2k-1)\pi}{2n}\right),$$

for $1 \le j, k \le n$, where δ_{jk} is the Kronecker delta. Given any arbitrary matrix $A_n \in \mathbb{C}^{n \times n}$, we define an operator Ψ_c which maps A_n to $\Psi_c(A_n)$ that minimizes $\|A_n - B_n\|_F$ over all $B_n \in \mathcal{C}$, see [18]. The construction of $\Psi_c(A_n)$ is also given in [18].

Optimal preconditioner based on the Hartley transform

Let $\mathcal{H}$ be the set of all $n \times n$ matrices that can be diagonalized by the discrete Hartley transform matrix Φ^h, i.e.,

$$\mathcal{H} = \{\Phi^h \Lambda_n \Phi^h \mid \Lambda_n \text{ is an } n \times n \text{ diagonal matrix}\}.$$

Here the (j, k)-th entry of Φ^h is given by

$$\frac{1}{\sqrt{n}} \cos\left(\frac{2\pi jk}{n}\right) + \frac{1}{\sqrt{n}} \sin\left(\frac{2\pi jk}{n}\right),$$

for $0 \leq j, k \leq n - 1$. Given any arbitrary matrix $A_n \in \mathbb{C}^{n \times n}$, we define an operator Ψ_h which maps A_n to $\Psi_h(A_n)$ that minimizes $\|A_n - B_n\|_F$ over all $B_n \in \mathcal{H}$, see [8]. The construction of $\Psi_h(A_n)$ is also given in [8].

Convergence result and operation cost

For solving Toeplitz systems $T_n u = b$ by using the PCG method with the optimal preconditioners based on fast transforms, we have the following theorem for the convergence rate, see [18, 25, 52].

Theorem 1.16 *Let $f > 0$ be an even function in $\mathcal{C}_{2\pi}$. Then the spectra of $(\Psi_\alpha(T_n))^{-1} T_n$ are clustered around 1 for large n, where $\alpha = s, c, h$. Thus the convergence rate of the PCG method is superlinear.*

In each iteration of the PCG method, see Section 1.1.4, we have to compute the matrix-vector multiplication $T_n v$ and solve the system

$$\Psi_\alpha(T_n) y = w.$$

We have already known that $T_n v$ can be computed in $O(n \log n)$ operations. The system $\Psi_\alpha(T_n) y = w$ can also be solved in $O(n \log n)$ operations by using the fast sine transform for $\Psi_\alpha(T_n) = \Psi_s(T_n) \in \mathcal{S}$; by using the fast cosine transform for $\Psi_\alpha(T_n) = \Psi_c(T_n) \in \mathcal{C}$; or by using the fast Hartley transform for $\Psi_\alpha(T_n) = \Psi_h(T_n) \in \mathcal{H}$. Thus, by Theorem 1.16, the complexity of algorithms remains $O(n \log n)$.

1.3.2 Band Toeplitz preconditioners

R. Chan and Tang proposed in [27] to use band Toeplitz matrices B_n as preconditioners for solving Toeplitz systems $T_n u = b$ by the PCG method, where T_n are assumed to be generated by a function $f \in C_{2\pi}$ with zeros. By Theorem 1.11, we know that T_n is ill conditioned. The generating function g of B_n is constructed not only to match the zeros in f but also to minimize

$$\left|\left|\left|\frac{f-g}{f}\right|\right|\right|_\infty$$

where $||| \cdot |||_\infty$ is the supremum norm. They showed that the condition number $\kappa(B_n^{-1}T_n)$ of the preconditioned matrix is $O(1)$.

Theorem 1.17 *Let $f \geq 0$ be the generating function of T_n and g be the generating function of a band Toeplitz matrix B_n:*

$$g(x) = \sum_{k=-N}^{N} b_k e^{ikx}$$

with $b_{-k} = \bar{b}_k$. If

$$\left|\left|\left|\frac{f-g}{f}\right|\right|\right|_\infty = h < 1,$$

then B_n is positive definite and

$$\kappa(B_n^{-1}T_n) \leq \frac{1+h}{1-h}$$

for all $n > 0$.

The function g could be found by a version of Remez algorithm. The algorithm also gives *a priori* bound on the number of iterations for convergence. To avoid the use of Remez algorithm, Serra proposed in [78] to use a technique of Chebyshev interpolation to find the function g. We refer to [80] for a comparison between the optimal preconditioners based on fast transforms and the band Toeplitz preconditioners.

1.3.3 $\{\omega\}$-circulant preconditioner

Potts and Steidl proposed in [71] to use $\{\omega\}$-circulant preconditioners to handle ill conditioned Toeplitz systems $T_n(f)u = b$. Here $f \in C_{2\pi}$ with zeros in $[-\pi, \pi]$. The preconditioner $P_n(f)$ is constructed as follows.

We choose uniform grids

$$x_k = w_n + \frac{2\pi k}{n}, \qquad w_n \in \left[-\pi, -\pi + \frac{2\pi}{n}\right)$$

such that

$$f(x_k) \neq 0,$$

for all $k = 0, \ldots, n-1$. Note that the choice of the grids requires some preparatory information about the zeros of f. Consider the preconditioner $P_n(f)$ that is of the form

$$P_n(f) = \Omega_n^* F_n^* \Lambda_n F_n \Omega_n \qquad (1.15)$$

where F_n is the Fourier matrix,

$$\Omega_n = \mathrm{diag}(1, e^{iw_n}, e^{i2w_n}, \cdots, e^{i(n-1)w_n})$$

and

$$\Lambda_n = \mathrm{diag}(f(x_0), f(x_1), f(x_2), \cdots, f(x_{n-1})).$$

The preconditioner $P_n(f)$ has the following properties, see [71]:

(i) The $P_n(f)$ is Hermitian positive definite if $f \geq 0$.

(ii) The $P_n(f)$ is a $\{e^{inw_n}\}$-circulant matrix, see [33]. Notice that $\{e^{inw_n}\}$-circulant matrix is Toeplitz matrix with the first entry of each column obtained by multiplying the last entry of the preceding column by e^{inw_n}.

(iii) Similarly to that of circulant matrices, once the diagonal matrix Λ_n in (1.15) is obtained, the products of $P_n(f)y$ and $P_n(f)^{-1}y$ for any vector y can be computed by FFTs in $O(n \log n)$ operations.

(iv) In view of (1.15), $P_n(f)$ can be constructed in $O(n \log n)$ operations and requires only $O(n)$ storage.

(v) The eigenvalues of $P_n(f)^{-1}T_n(f)$ are clustered around 1 and the spectra of $P_n(f)^{-1}T_n(f)$ are bounded away from zero.

The PCG method, when applied to solving the preconditioned system with the preconditioner $P_n(f)$, will converge superlinearly. Therefore, the total complexity in solving the preconditioned system remains $O(n \log n)$.

Chapter 2

Block Circulant Preconditioners

We study the solution of block system

$$T_{mn}u = b$$

by the PCG method where

$$
T_{mn} = \begin{pmatrix}
T_{(0)} & T_{(-1)} & \cdots & T_{(2-m)} & T_{(1-m)} \\
T_{(1)} & T_{(0)} & T_{(-1)} & \cdots & T_{(2-m)} \\
\vdots & T_{(1)} & T_{(0)} & \ddots & \vdots \\
T_{(m-2)} & \cdots & \ddots & \ddots & T_{(-1)} \\
T_{(m-1)} & T_{(m-2)} & \cdots & T_{(1)} & T_{(0)}
\end{pmatrix}
\tag{2.1}
$$

and the blocks $T_{(l)}$, for $l = 0, \pm 1, \pm 2, \cdots, \pm(m-1)$, are themselves Toeplitz matrices of order n. The matrix T_{mn} is a block Toeplitz matrix with Toeplitz blocks, and will be called the BTTB matrix. Several preconditioners that preserve the block structure of T_{mn} are constructed. We show that they are good preconditioners for solving some BTTB systems. Since our block preconditioners are defined not just for T_{mn} given by (2.1) but for general matrices as well, we then begin with the general case.

2.1 Operators for block matrices

In the following, we call c_U, defined in (i) of Lemma 1.1, the point operator in order to distinguish it from the block operators that we now introduce. Let us begin by considering a general block system $A_{mn}x = b$ where A_{mn} is partitioned as follows,

$$A_{mn} = \begin{pmatrix} A_{1,1} & A_{1,2} & \cdots & A_{1,m} \\ A_{2,1} & A_{2,2} & \cdots & A_{2,m} \\ \vdots & \ddots & \ddots & \vdots \\ A_{m,1} & A_{m,2} & \cdots & A_{m,m} \end{pmatrix} \tag{2.2}$$

with $A_{i,j} \in \mathbb{C}^{n \times n}$.

2.1.1 Block operator $c_U^{(1)}$

Since we are interested in solving block systems where the blocks are Toeplitz matrices, in view of the point case, a natural choice of preconditioner for A_{mn} given by (2.2) is

$$E_{mn} \equiv \begin{pmatrix} c_F(A_{1,1}) & c_F(A_{1,2}) & \cdots & c_F(A_{1,m}) \\ c_F(A_{2,1}) & c_F(A_{2,2}) & \cdots & c_F(A_{2,m}) \\ \vdots & & \ddots & \ddots & \vdots \\ c_F(A_{m,1}) & c_F(A_{m,2}) & \cdots & c_F(A_{m,m}) \end{pmatrix}$$

where the blocks $c_F(A_{i,j})$ are just the point circulant approximations to $A_{i,j}$, see (1.10). We will show in Sections 2.3 and 2.5 that E_{mn} is a good preconditioner for solving some block systems.

Now, we are going to study the spectral properties. Let $\delta^{(1)}(A_{mn})$ be defined by

$$\delta^{(1)}(A_{mn}) \equiv \begin{pmatrix} \delta(A_{1,1}) & \delta(A_{1,2}) & \cdots & \delta(A_{1,m}) \\ \delta(A_{2,1}) & \delta(A_{2,2}) & \cdots & \delta(A_{2,m}) \\ \vdots & & \ddots & \ddots & \vdots \\ \delta(A_{m,1}) & \delta(A_{m,2}) & \cdots & \delta(A_{m,m}) \end{pmatrix} \tag{2.3}$$

where each block $\delta(A_{i,j})$, defined as in Section 1.2.2, is a diagonal matrix whose diagonal is equal to the diagonal of the matrix $A_{i,j}$. The following lemma gives the relation between $\sigma_{\max}(A_{mn})$ and $\sigma_{\max}(\delta^{(1)}(A_{mn}))$ where $\sigma_{\max}(\cdot)$ denotes the largest singular value.

Lemma 2.1 *For any arbitrary $A_{mn} \in \mathbb{C}^{mn \times mn}$ partitioned as in (2.2), we have*

$$\sigma_{\max}(\delta^{(1)}(A_{mn})) \leq \sigma_{\max}(A_{mn}). \tag{2.4}$$

Furthermore, when A_{mn} is Hermitian, we have

$$\lambda_{\min}(A_{mn}) \leq \lambda_{\min}(\delta^{(1)}(A_{mn})) \leq \lambda_{\max}(\delta^{(1)}(A_{mn})) \leq \lambda_{\max}(A_{mn}). \tag{2.5}$$

In particular, if A_{mn} is positive definite, then so is $\delta^{(1)}(A_{mn})$.

Proof: Let

$$(A_{mn})_{i,j;k,l} = (A_{k,l})_{ij}$$

be the (i,j)-th entry of the (k,l)-th block of A_{mn}. Let P be the permutation matrix which satisfies

$$(P^* A_{mn} P)_{k,l;i,j} = (A_{mn})_{i,j;k,l}, \tag{2.6}$$

for $1 \leq i,j \leq n, 1 \leq k,l \leq m$, and let

$$B_{mn} \equiv P^* \delta^{(1)}(A_{mn}) P.$$

Then B_{mn} is of the following form

$$B_{mn} = \begin{pmatrix} B_{1,1} & 0 & \cdots & 0 \\ 0 & B_{2,2} & \cdots & 0 \\ \vdots & & \ddots & \vdots \\ 0 & 0 & \cdots & B_{n,n} \end{pmatrix}.$$

We know that B_{mn} and $\delta^{(1)}(A_{mn})$ have the same singular values and eigenvalues. For each k, since $B_{k,k}$ is a principal submatrix of the matrix A_{mn}, it follows that

$$\sigma_{\max}(B_{k,k}) \leq \sigma_{\max}(A_{mn}),$$

see for instance, [92]. Hence we have

$$\sigma_{\max}(\delta^{(1)}(A_{mn})) = \sigma_{\max}(B_{mn}) = \max_k (\sigma_{\max}(B_{k,k})) \leq \sigma_{\max}(A_{mn}).$$

When A_{mn} is Hermitian, by Cauchy's Interlace Theorem, we then have

$$\lambda_{\min}(A_{mn}) \le \min_k \left(\lambda_{\min}(B_{k,k})\right) = \lambda_{\min}(\delta^{(1)}(A_{mn}))$$

$$\le \lambda_{\max}(\delta^{(1)}(A_{mn})) = \max_k \left(\lambda_{\max}(B_{k,k})\right)$$

$$\le \lambda_{\max}(A_{mn}). \quad \square$$

In the following, let $\mathcal{D}_{m,n}^{(1)}$ denote the set of all matrices of the form given by (2.3), i.e., $\mathcal{D}_{m,n}^{(1)}$ is the set of all $m \times m$ block matrices with $n \times n$ diagonal blocks and let

$$\mathcal{M}_U^{(1)} = \{(I \otimes U)^* \Lambda_{mn}^{(1)} (I \otimes U) \mid \Lambda_{mn}^{(1)} \in \mathcal{D}_{m,n}^{(1)}\}$$

where I is the $m \times m$ identity matrix and U is any given $n \times n$ unitary matrix. We then define an operator $c_U^{(1)}$ which maps every $A_{mn} \in \mathbb{C}^{mn \times mn}$ to the minimizer of $\|W_{mn} - A_{mn}\|_F$ over all $W_{mn} \in \mathcal{M}_U^{(1)}$. Some properties of this operator are given in the following theorem.

Theorem 2.1 *For any arbitrary $A_{mn} \in \mathbb{C}^{mn \times mn}$ partitioned as in (2.2), let $c_U^{(1)}(A_{mn})$ be the minimizer of $\|W_{mn} - A_{mn}\|_F$ over all $W_{mn} \in \mathcal{M}_U^{(1)}$. Then*

(i) *$c_U^{(1)}(A_{mn})$ is uniquely determined by A_{mn} and is given by*

$$c_U^{(1)}(A_{mn}) = (I \otimes U)^* \delta^{(1)}[(I \otimes U)A_{mn}(I \otimes U)^*](I \otimes U). \quad (2.7)$$

(ii) *$c_U^{(1)}(A_{mn})$ is also given by*

$$c_U^{(1)}(A_{mn}) = \begin{pmatrix} c_U(A_{1,1}) & c_U(A_{1,2}) & \cdots & c_U(A_{1,m}) \\ c_U(A_{2,1}) & c_U(A_{2,2}) & \cdots & c_U(A_{2,m}) \\ \vdots & \ddots & \ddots & \vdots \\ c_U(A_{m,1}) & c_U(A_{m,2}) & \cdots & c_U(A_{m,m}) \end{pmatrix} \quad (2.8)$$

where c_U is the point operator defined by (1.9).

(iii) *We have*

$$\sigma_{\max}(c_U^{(1)}(A_{mn})) \le \sigma_{\max}(A_{mn}). \qquad (2.9)$$

(iv) *If A_{mn} is Hermitian, then $c_U^{(1)}(A_{mn})$ is also Hermitian and*

$$\lambda_{\min}(A_{mn}) \le \lambda_{\min}(c_U^{(1)}(A_{mn})) \le \lambda_{\max}(c_U^{(1)}(A_{mn})) \le \lambda_{\max}(A_{mn}).$$

In particular, if A_{mn} is positive definite, then so is $c_U^{(1)}(A_{mn})$.

(v) *The operator $c_U^{(1)}$ is a linear projection operator from $\mathbb{C}^{mn \times mn}$ into $\mathcal{M}_U^{(1)}$ and has the operator norms*

$$\|c_U^{(1)}\|_2 \equiv \sup_{\|A_{mn}\|_2 = 1} \|c_U^{(1)}(A_{mn})\|_2 = 1$$

and

$$\|c_U^{(1)}\|_F \equiv \sup_{\|A_{mn}\|_F = 1} \|c_U^{(1)}(A_{mn})\|_F = 1.$$

Proof:

(i) Let $W_{mn} \in \mathcal{M}_U^{(1)}$ be given by

$$W_{mn} = (I \otimes U)^* \Lambda_{mn}^{(1)} (I \otimes U)$$

where $\Lambda_{mn}^{(1)} \in \mathcal{D}_{m,n}^{(1)}$. Since the Frobenius norm is unitary invariant, see Section 1.1.1, we have

$$\|W_{mn} - A_{mn}\|_F = \|(I \otimes U)^* \Lambda_{mn}^{(1)} (I \otimes U) - A_{mn}\|_F$$

$$= \|\Lambda_{mn}^{(1)} - (I \otimes U) A_{mn} (I \otimes U)^*\|_F.$$

Thus, the minimizing problem $\|W_{mn} - A_{mn}\|_F$ over $\mathcal{M}_U^{(1)}$ is equivalent to the minimizing problem

$$\|\Lambda_{mn}^{(1)} - (I \otimes U) A_{mn} (I \otimes U)^*\|_F$$

over $\mathcal{D}_{m,n}^{(1)}$. Since $\Lambda_{mn}^{(1)}$ can only affect the diagonal of each block of

$$(I \otimes U) A_{mn} (I \otimes U)^*,$$

the solution for the latter problem is

$$\Lambda_{mn}^{(1)} = \delta^{(1)}[(I \otimes U) A_{mn} (I \otimes U)^*].$$

Hence

$$c_U^{(1)}(A_{mn}) = (I \otimes U)^* \delta^{(1)}[(I \otimes U)A_{mn}(I \otimes U)^*](I \otimes U)$$

is the minimizer of $\|W_{mn} - A_{mn}\|_F$ over all $W_{mn} \in \mathcal{M}_U^{(1)}$. Obviously, $\Lambda_{mn}^{(1)}$ and hence $c_U^{(1)}(A_{mn})$ are uniquely determined by A_{mn}.

(ii) Since

$$\delta^{(1)}[(I \otimes U)A_{mn}(I \otimes U)^*] =$$

$$\begin{pmatrix} \delta(UA_{1,1}U^*) & \delta(UA_{1,2}U^*) & \cdots & \delta(UA_{1,m}U^*) \\ \delta(UA_{2,1}U^*) & \delta(UA_{2,2}U^*) & \cdots & \delta(UA_{2,m}U^*) \\ \vdots & \ddots & \ddots & \vdots \\ \delta(UA_{m,1}U^*) & \delta(UA_{m,2}U^*) & \cdots & \delta(UA_{m,m}U^*) \end{pmatrix},$$

by (1.9) and (2.7), we see that (2.8) holds.

(iii) For any $A_{mn} \in \mathbb{C}^{mn \times mn}$, we have by (2.7) and (2.4)

$$\sigma_{\max}(c_U^{(1)}(A_{mn})) = \sigma_{\max}[\delta^{(1)}((I \otimes U)A_{mn}(I \otimes U)^*)]$$

$$\leq \sigma_{\max}[(I \otimes U)A_{mn}(I \otimes U)^*] = \sigma_{\max}(A_{mn}).$$

(iv) If A_{mn} is Hermitian, then it is clear from (2.8) and Lemma 1.1 (ii) that $c_U^{(1)}(A_{mn})$ is also Hermitian. Moreover, by (2.5) and (2.7), we have

$$\lambda_{\min}(A_{mn}) = \lambda_{\min}[(I \otimes U)A_{mn}(I \otimes U)^*]$$

$$\leq \lambda_{\min}[\delta^{(1)}((I \otimes U)A_{mn}(I \otimes U)^*)]$$

$$= \lambda_{\min}(c_U^{(1)}(A_{mn})) \leq \lambda_{\max}(c_U^{(1)}(A_{mn}))$$

$$= \lambda_{\max}[\delta^{(1)}((I \otimes U)A_{mn}(I \otimes U)^*)]$$

$$\leq \lambda_{\max}[(I \otimes U)A_{mn}(I \otimes U)^*] = \lambda_{\max}(A_{mn}).$$

(v) By (2.9), we have

$$\|c_U^{(1)}(A_{mn})\|_2 = \sigma_{\max}[c_U^{(1)}(A_{mn})] \leq \sigma_{\max}(A_{mn}) = \|A_{mn}\|_2.$$

However, for the identity matrix I_{mn}, we have

$$\|c_U^{(1)}(I_{mn})\|_2 = \|I_{mn}\|_2 = 1.$$

Hence $\|c_U^{(1)}\|_2 = 1$. For the Frobenius norm, since

$$\|c_U^{(1)}(A_{mn})\|_F = \|\delta^{(1)}[(I \otimes U)A_{mn}(I \otimes U)^*]\|_F$$

$$\leq \|(I \otimes U)A_{mn}(I \otimes U)^*\|_F = \|A_{mn}\|_F$$

and

$$\|c_U^{(1)}(\frac{1}{\sqrt{mn}}I_{mn})\|_F = \frac{1}{\sqrt{mn}}\|I_{mn}\|_F = 1,$$

it follows that $\|c_U^{(1)}\|_F = 1$. $\square$

2.1.2 Block operator $\tilde{c}_V^{(1)}$

For matrices A_{mn} partitioned as in (2.2), we can define another block approximation to them. Let $\tilde{\delta}^{(1)}(A_{mn})$ be defined by

$$\tilde{\delta}^{(1)}(A_{mn}) \equiv \begin{pmatrix} A_{1,1} & 0 & \cdots & 0 \\ 0 & A_{2,2} & \cdots & 0 \\ \vdots & \ddots & \ddots & \vdots \\ 0 & \cdots & 0 & A_{m,m} \end{pmatrix}. \tag{2.10}$$

In the following, we use $\widetilde{\mathcal{D}}_{m,n}^{(1)}$ to denote the set of all matrices of the form given by (2.10), i.e., $\widetilde{\mathcal{D}}_{m,n}^{(1)}$ is the set of all $m \times m$ block diagonal matrices with $n \times n$ blocks. Let

$$\widetilde{\mathcal{M}}_V^{(1)} = \{(V \otimes I)^* \tilde{\Lambda}_{mn}^{(1)} (V \otimes I) \mid \tilde{\Lambda}_{mn}^{(1)} \in \widetilde{\mathcal{D}}_{m,n}^{(1)}\}$$

where V is any given $m \times m$ unitary matrix and I is the $n \times n$ identity matrix.

We define an operator $\tilde{c}_V^{(1)}$ that maps every $A_{mn} \in \mathbb{C}^{mn \times mn}$ to the minimizer of $\|W_{mn} - A_{mn}\|_F$ over all $W_{mn} \in \widetilde{\mathcal{M}}_V^{(1)}$. Similarly to Theorem 2.1, we have the following theorem.

Theorem 2.2 *For any arbitrary $A_{mn} \in \mathbb{C}^{mn \times mn}$ partitioned as in (2.2), let $\tilde{c}_V^{(1)}(A_{mn})$ be the minimizer of $\|W_{mn} - A_{mn}\|_F$ over all $W_{mn} \in \widetilde{\mathcal{M}}_V^{(1)}$. Then*

(i) $\tilde{c}_V^{(1)}(A_{mn})$ *is uniquely determined by A_{mn} and is given by*

$$\tilde{c}_V^{(1)}(A_{mn}) = (V \otimes I)^* \tilde{\delta}^{(1)}[(V \otimes I)A_{mn}(V \otimes I)^*](V \otimes I). \quad (2.11)$$

(ii) *We have*

$$\sigma_{\max}(\tilde{c}_V^{(1)}(A_{mn})) \leq \sigma_{\max}(A_{mn}).$$

(iii) *If A_{mn} is Hermitian, then $\tilde{c}_V^{(1)}(A_{mn})$ is also Hermitian and*

$$\lambda_{\min}(A_{mn}) \leq \lambda_{\min}(\tilde{c}_V^{(1)}(A_{mn})) \leq \lambda_{\max}(\tilde{c}_V^{(1)}(A_{mn})) \leq \lambda_{\max}(A_{mn}).$$

In particular, if A_{mn} is positive definite then so is $\tilde{c}_V^{(1)}(A_{mn})$.

(iv) *The operator $\tilde{c}_V^{(1)}$ is a linear projection operator from $\mathbb{C}^{mn \times mn}$ into $\widetilde{\mathcal{M}}_V^{(1)}$ and has the operator norms*

$$\|\tilde{c}_V^{(1)}\|_2 = \|\tilde{c}_V^{(1)}\|_F = 1.$$

We omit the proof of Theorem 2.2 since it is quite similar to that of Theorem 2.1. However, we note that (ii)–(iv) in Theorem 2.2 can be proved easily by using the following relationship between $c_U^{(1)}$ and $\tilde{c}_V^{(1)}$.

Lemma 2.2 *Let U be any given unitary matrix and P be the permutation matrix defined as in (2.6). For any arbitrary matrix $A_{mn} \in \mathbb{C}^{mn \times mn}$ partitioned as in (2.2), we have*

$$\delta^{(1)}(A_{mn}) = P \tilde{\delta}^{(1)}(P^* A_{mn} P) P^*$$

and

$$c_U^{(1)}(A_{mn}) = P \tilde{c}_U^{(1)}(P^* A_{mn} P) P^*.$$

Proof: To prove the first equality by the definition of $\tilde{\delta}^{(1)}$ and (2.6), we note that

$$[\tilde{\delta}^{(1)}(P^*A_{mn}P)]_{k,l;i,j} = \begin{cases} (P^*A_{mn}P)_{k,l;i,j}, & i = j, \\ \\ 0, & i \neq j, \end{cases}$$

$$= \begin{cases} (A_{mn})_{i,j;k,l}, & i = j, \\ \\ 0, & i \neq j. \end{cases}$$

Hence

$$[P\tilde{\delta}^{(1)}(P^*A_{mn}P)P^*]_{i,j;k,l}$$

$$= [\tilde{\delta}^{(1)}(P^*A_{mn}P)]_{k,l;i,j} = \begin{cases} (A_{mn})_{i,j;k,l}, & i = j, \\ \\ 0, & i \neq j, \end{cases}$$

which by definition is equal to $[\delta^{(1)}(A_{mn})]_{i,j;k,l}$. To prove the second equality, since

$$(I \otimes U)P = P(U \otimes I)$$

for any matrix U, we have by (2.11) and (2.7),

$$P\tilde{c}_U^{(1)}(P^*A_{mn}P)P^* = P(U \otimes I)^*\tilde{\delta}^{(1)}[(U \otimes I)P^*A_{mn}P(U \otimes I)^*](U \otimes I)P^*$$

$$= (I \otimes U)^*P\tilde{\delta}^{(1)}[P^*(I \otimes U)A_{mn}(I \otimes U)^*P]P^*(I \otimes U)$$

$$= (I \otimes U)^*\delta^{(1)}[(I \otimes U)A_{mn}(I \otimes U)^*](I \otimes U)$$

$$= c_U^{(1)}(A_{mn}). \quad \square$$

2.1.3   Operator $c_{V,U}^{(2)}$

Intuitively, $c_U^{(1)}(A_{mn})$ and $\tilde{c}_V^{(1)}(A_{mn})$ resemble the diagonalization of A_{mn} along one specific direction. It is natural to consider the matrix that results from diagonalization along both directions. Let

$$c_{V,U}^{(2)} \equiv \tilde{c}_V^{(1)} \circ c_U^{(1)}$$

where 'o' denotes the composite of the operators. The following lemma derives the properties of operator $c_{V,U}^{(2)}$.

Lemma 2.3 *For any arbitrary matrix $A_{mn} \in \mathbb{C}^{mn \times mn}$ partitioned as in (2.2), we have*

$$(I \otimes U)^* \tilde{\delta}^{(1)}(A_{mn})(I \otimes U) = \tilde{\delta}^{(1)}[(I \otimes U)^* A_{mn}(I \otimes U)] \qquad (2.12)$$

and

$$(V \otimes I)\delta^{(1)}(A_{mn})(V \otimes I)^* = \delta^{(1)}[(V \otimes I)A_{mn}(V \otimes I)^*]. \qquad (2.13)$$

Furthermore,

$$\tilde{\delta}^{(1)} \circ \delta^{(1)}(A_{mn}) = \delta(A_{mn}) = \delta^{(1)} \circ \tilde{\delta}^{(1)}(A_{mn}). \qquad (2.14)$$

The proof of Lemma 2.3 is straightforward, we therefore omit it. Let

$$\mathcal{M}_{V \otimes U} = \{(V \otimes U)^* \Lambda_{mn}(V \otimes U) \,|\, \Lambda_{mn} \text{ is an } mn \times mn \text{ diagonal matrix}\}$$

where V is any given $m \times m$ unitary matrix and U is any given $n \times n$ unitary matrix. We can prove the following theorem by using Lemma 2.3.

Theorem 2.3 *For any arbitrary matrix $A_{mn} \in \mathbb{C}^{mn \times mn}$ partitioned as in (2.2), let $c_{V \otimes U}(A_{mn})$ be the minimizer of $\|W_{mn} - A_{mn}\|_F$ over all $W_{mn} \in \mathcal{M}_{V \otimes U}$ where $c_{V \otimes U}$ is the point operator defined as in Lemma 1.1. Then*

(i) *We have*
$$c_{V,U}^{(2)}(A_{mn}) = c_{V \otimes U}(A_{mn}).$$

(ii) *If A_{mn} is Hermitian then $c_{V,U}^{(2)}(A_{mn})$ is also Hermitian and*

$$\lambda_{\min}(A_{mn}) \le \lambda_{\min}(c_{V,U}^{(2)}(A_{mn})) \le \lambda_{\max}(c_{V,U}^{(2)}(A_{mn})) \le \lambda_{\max}(A_{mn}).$$

In particular, if A_{mn} is positive definite, then so is $c_{V,U}^{(2)}(A_{mn})$.

(iii) *The operator $c_{V,U}^{(2)}$ has the operator norms*

$$\|c_{V,U}^{(2)}\|_2 = \|c_{V,U}^{(2)}\|_F = 1.$$

Proof: We only prove (i). It is straightforward to show (ii) and (iii) by using Lemma 1.1 with the point operator $c_{V \otimes U}$. For any given A_{mn}, by definitions of $c_U^{(1)}$ and $\tilde{c}_V^{(1)}$, we have

$$c_{V,U}^{(2)}(A_{mn}) = \tilde{c}_V^{(1)}[c_U^{(1)}(A_{mn})]$$

$$= (V \otimes I)^* \tilde{\delta}^{(1)} \{(V \otimes I)[(I \otimes U)^* \delta^{(1)}[(I \otimes U)A_{mn}(I \otimes U)^*]$$

$$\times (I \otimes U)](V \otimes I)^*\}(V \otimes I)$$

$$= (V \otimes I)^* \tilde{\delta}^{(1)} \{(I \otimes U)^* (V \otimes I) \delta^{(1)}[(I \otimes U)A_{mn}(I \otimes U)^*]$$

$$\times (V \otimes I)^* (I \otimes U)\}(V \otimes I).$$

Hence by (2.12), (2.13) and (2.14), we have

$$c_{V,U}^{(2)}(A_{mn}) = (V \otimes U)^* \tilde{\delta}^{(1)} \{[\delta^{(1)}[(V \otimes U)A_{mn}(V \otimes U)^*]\}(V \otimes U)$$

$$= (V \otimes U)^* \delta[(V \otimes U)A_{mn}(V \otimes U)^*](V \otimes U)$$

$$= c_{V \otimes U}(A_{mn}). \quad \square$$

We remark that $c_{V,U}^{(2)}(A_{mn})$ is an approximation of A_{mn} in two directions.

2.1.4 Three formulae

When U and V take the Fourier matrix F, we give three simple formulae for finding $c_F^{(1)}(A_{mn})$, $\tilde{c}_F^{(1)}(A_{mn})$ and $c_{F,F}^{(2)}(A_{mn})$. We have by (2.8),

$$c_F^{(1)}(A_{mn}) = \begin{pmatrix} c_F(A_{1,1}) & c_F(A_{1,2}) & \cdots & c_F(A_{1,m}) \\ c_F(A_{2,1}) & c_F(A_{2,2}) & \cdots & c_F(A_{2,m}) \\ \vdots & \ddots & \ddots & \vdots \\ c_F(A_{m,1}) & c_F(A_{m,2}) & \cdots & c_F(A_{m,m}) \end{pmatrix}, \qquad (2.15)$$

where each block $c_F(A_{i,j})$ is T. Chan's circulant preconditioner for $A_{i,j}$.

Next we find $\tilde{c}_F^{(1)}(A_{mn})$ by using Lemma 2.2. Let $A_{mn} = P^* B_{mn} P$ where P is defined as in (2.6). Then B_{mn} is partitioned into n^2 blocks with each block $B_{i,j} \in \mathbb{C}^{m \times m}$. By Lemma 2.2 and (2.15), we have

$$[\tilde{c}_F^{(1)}(A_{mn})]_{i,j;k,l} = [P^* c_F^{(1)}(B_{mn}) P]_{i,j;k,l} = [c_F^{(1)}(B_{mn})]_{k,l;i,j} = (c_F(B_{i,j}))_{kl}$$

where $B_{i,j}$ is the (i,j)-th block of the matrix B_{mn}. By (1.10), we see that the (k,l)-th entry of the circulant matrix $c_F(B_{i,j})$ is given by

$$(c_F(B_{i,j}))_{kl} = \frac{1}{m} \sum_{p-q \equiv k-l (\bmod m)} (B_{i,j})_{pq}.$$

Since $(B_{i,j})_{pq} = (A_{p,q})_{ij}$, we have

$$[\tilde{c}_F^{(1)}(A_{mn})]_{i,j;k,l} = \frac{1}{m} \sum_{p-q \equiv k-l (\bmod m)} (A_{p,q})_{ij},$$

for $1 \leq i,j \leq n, 1 \leq k,l \leq m$. Thus, the (k,l)-th block of $\tilde{c}_F^{(1)}(A_{mn})$ is given by

$$\frac{1}{m} \sum_{p-q \equiv k-l (\bmod m)} (A_{pq}).$$

Since it depends only on $k - l (\bmod m)$, we see that $\tilde{c}_F^{(1)}(A_{mn})$ is a block circulant matrix. Using the matrix Q defined as in (1.11), we further have

$$\tilde{c}_F^{(1)}(A_{mn}) = \frac{1}{m} \sum_{j=0}^{m-1} \left(Q^j \otimes \sum_{p-q \equiv j (\bmod m)} A_{p,q} \right). \qquad (2.16)$$

Finally, by using (1.10), (2.16) and Theorem 2.3, one can easily obtain the following formula,

$$c_{F,F}^{(2)}(A_{mn}) = \frac{1}{mn} \sum_{j=0}^{m-1} \sum_{k=0}^{n-1} \left(\sum_{p-q \equiv j (\bmod m)} \sum_{r-s \equiv k (\bmod n)} (A_{p,q})_{s,t} \right) (Q^j \otimes Q^k). \qquad (2.17)$$

We remark that $c_{F,F}^{(2)}(A_{mn})$ is a block circulant matrix with circulant blocks, see [33], and will be called the BCCB matrix. Actually, from Theorem 2.3, we know that $c_{F,F}^{(2)}(A_{mn})$ is the minimizer of $\|W_{mn} - A_{mn}\|_F$ over all $W_{mn} \in \mathcal{M}_{F \otimes F}$ where $\mathcal{M}_{F \otimes F}$ is the set of all BCCB matrices.

2.2 Operation cost for preconditioned system

In this section, we consider the cost of solving block systems $A_{mn}x = b$ by the PCG method with preconditioners $c_F^{(1)}(A_{mn})$ and $c_{F,F}^{(2)}(A_{mn})$. The analysis for $\tilde{c}_F^{(1)}(A_{mn})$ is similar. We first recall that in each iteration of the PCG method, we have to compute the matrix-vector multiplication $A_{mn}v$ for some vector v and solve the systems

$$c_F^{(1)}(A_{mn})y = d \qquad (2.18)$$

or

$$c_{F,F}^{(2)}(A_{mn})y = d \qquad (2.19)$$

for some vector d, see Section 1.1.4 and [38, 75].

2.2.1 General matrices

Let $A_{mn} \in \mathbb{C}^{mn \times mn}$ be a general matrix.

Case of $c_F^{(1)}(A_{mn})$

We note that by (2.7), the solution of (2.18) is given by

$$y = (I \otimes F)^*[\delta^{(1)}((I \otimes F)A_{mn}(I \otimes F)^*)]^{-1}(I \otimes F)d. \qquad (2.20)$$

Hence before we start the iteration, we should form the matrix

$$\Delta \equiv \delta^{(1)}((I \otimes F)A_{mn}(I \otimes F)^*)$$

and compute its inverse. We note that by (2.15), the (i, j)-th block of Δ is just $Fc_F(A_{i,j})F^*$. By (1.9),

$$Fc_F(A_{i,j})F^* = \delta(FA_{i,j}F^*)$$

and hence can be computed in n^2 operations and one FFT, see [21]. Thus the cost of obtaining Δ is $O(m^2n^2)$ operations. Next we compute its inverse. We first permute the matrix Δ by P defined as in (2.6) to obtain

$$B_{mn} \equiv P^*\Delta P = \begin{pmatrix} B_{1,1} & 0 & \cdots & 0 \\ 0 & B_{2,2} & \cdots & 0 \\ \vdots & \ddots & \ddots & \vdots \\ 0 & 0 & \cdots & B_{n,n} \end{pmatrix}.$$

We then compute the LU decompositions for all diagonal blocks $B_{k,k}$. That will take $O(nm^3)$ operations. In total it requires $O(m^2n^2 + nm^3)$ operations in the initialization step.

After obtaining the LU factors of Δ, we start the iteration. For a general dense matrix A_{mn}, $A_{mn}v$ can be computed in $O(m^2n^2)$. To get the vector y in (2.20), we note that the vectors of the form $(I \otimes F)d$ can be computed in $O(mn \log n)$ operations by using FFTs. By using the LU factors of Δ, $O(nm^2)$ operations are need to compute $\Delta^{-1}d$ for any vector d. Totally, the cost per iteration to get y is $O(mn \log n) + O(nm^2)$ operations.

We would like to mention that some of the block operations mentioned above can be done in parallel. For instance, the diagonal $\delta(FA_{i,j}F^*)$ of the blocks $c_F(A_{i,j})$ can be obtained in $O(n^2)$ parallel steps with $O(m^2)$ processors and the LU decompositions of the blocks $B_{k,k}$ in B_{mn} can also be computed in parallel. This can further reduce the cost per iteration.

Case of $c^{(2)}_{F,F}(A_{mn})$

For the solution of (2.19), we note that any BCCB matrix $C_{mn} \in \mathcal{M}_{F_m \otimes F_n}$ could be defined by its first column. We have the following relation between the first column and the eigenvalues of C_{mn},

$$\sqrt{mn}C_{mn}e_1 = (F_m^* \otimes F_n^*)\Lambda_{mn}1_{mn} \tag{2.21}$$

where

$$e_1 = (1, 0, \cdots, 0)^T \in \mathbb{R}^{mn}$$

is the first unit vector,

$$1_{mn} = (1, 1, \cdots, 1)^T$$

is the vector of all 1's, and Λ_{mn} is a diagonal matrix holding the eigenvalues of C_{mn}. Let $c_q^{(p)}$ denote the q-th entry in the p-th block of the first column of C_{mn}, for $0 \leq q \leq n-1$ and $0 \leq p \leq m-1$. Then

$$C_{mn}e_1 = (c_0^{(0)}, c_1^{(0)}, \cdots, c_{n-1}^{(0)}, c_0^{(1)}, \cdots, c_0^{(m-1)}, \cdots, c_{n-1}^{(m-1)})^T$$

is the first column of C_{mn}. We have

$$\lambda_{p,q}(C_{mn}) = (\Lambda_{mn})_{p,q} = \sum_{s=0}^{m-1}\sum_{t=0}^{n-1} c_t^{(s)}\xi_p^s\eta_q^t, \tag{2.22}$$

for $0 \le p \le m - 1$ and $0 \le q \le n - 1$, where

$$\xi_p = e^{\frac{2\pi i p}{m}} \quad \text{and} \quad \eta_q = e^{\frac{2\pi i q}{n}}$$

with $i \equiv \sqrt{-1}$. Note that

$$\xi_p^{m-j} = \xi_p^{-j} = \bar{\xi}_p^{j}$$

and

$$\eta_q^{n-k} = \eta_q^{-k} = \bar{\eta}_q^{k}.$$

By formula (2.17), $O(m^2 n^2)$ operations are required to compute the first column of $c_{F,F}^{(2)}(A_{mn})$. Followed by using (2.21) and the 2-dimensional FFT to compute the eigenvalues of $c_{F,F}^{(2)}(A_{mn})$, it requires $O(mn \log mn)$ operations. Thus $O(m^2 n^2)$ operations are required in the initialization step.

In each iteration step, besides the $O(m^2 n^2)$ operations to compute $A_{mn} v$, we need to solve (2.19). Since (2.19) could be changed as follows,

$$y = (F_m^* \otimes F_n^*) \Delta^{-1} (F_m \otimes F_n) d$$

where Δ is the diagonal matrix holding the eigenvalues of $c_{F,F}^{(2)}(A_{mn})$, by using 2-dimensional FFTs again, it requires $O(mn \log mn)$ operations to obtain y per iteration.

2.2.2   Level-2 symmetric BTTB matrices

Let us consider a family of BTTB systems $T_{mn} u = b$ where $T_{mn} \in \mathbb{R}^{mn \times mn}$ is of the form

$$T_{mn} = \begin{pmatrix} T_{1,1} & T_{1,2} & \cdots & T_{1,m} \\ T_{2,1} & T_{2,2} & \cdots & T_{2,m} \\ \vdots & \ddots & \ddots & \vdots \\ T_{m,1} & T_{m,2} & \cdots & T_{m,m} \end{pmatrix}$$

$$= \begin{pmatrix} T_{(0)} & T_{(1)} & \cdots & T_{(m-1)} \\ T_{(1)} & T_{(0)} & \cdots & T_{(m-2)} \\ \vdots & \ddots & \ddots & \vdots \\ T_{(m-1)} & T_{(m-2)} & \cdots & T_{(0)} \end{pmatrix}. \tag{2.23}$$

Here the blocks $T_{i,j} = T_{(|i-j|)} \in \mathbb{R}^{n \times n}$ are themselves symmetric Toeplitz matrices. Such T_{mn} are called level-2 symmetric BTTB matrices. We consider the cost of solving BTTB systems $T_{mn}u = b$ by the PCG method with preconditioners $c_F^{(1)}(T_{mn})$ and $c_{F,F}^{(2)}(T_{mn})$.

Case of $c_F^{(1)}(T_{mn})$

For $c_F^{(1)}(T_{mn})$, by (2.15), the blocks of it are just $c_F(T_{(k)})$. Hence by (1.10) and the fact that $T_{(k)}$ is Toeplitz, the diagonal $\delta(FT_{(k)}F^*)$ can be computed in $O(n \log n)$ operations. Therefore, we need $O(mn \log n)$ operations to form

$$\Delta = \delta^{(1)}((I \otimes F)T_{mn}(I \otimes F)^*).$$

We should emphasize that in this case, there is no need to compute the LU factors of Δ. In fact,

$$P^* \Delta P = \begin{pmatrix} \tilde{T}_{1,1} & 0 & \cdots & 0 \\ 0 & \tilde{T}_{2,2} & \cdots & 0 \\ \vdots & \ddots & \ddots & \vdots \\ 0 & 0 & \cdots & \tilde{T}_{n,n} \end{pmatrix}$$

where P is defined as in (2.6) and

$$(\tilde{T}_{k,k})_{ij} = (\delta(FT_{i,j}F^*))_{kk} = (\delta(FT_{(|i-j|)}F^*))_{kk},$$

for $1 \leq i, j \leq m$ and $1 \leq k \leq n$. Hence we see that the diagonal blocks $\tilde{T}_{k,k} \in \mathbb{R}^{m \times m}$ are still symmetric Toeplitz matrices. Therefore it requires only $O(m \log^2 m)$ operations to compute $\tilde{T}_{k,k}^{-1}v$ for any vector v, see Ammar and Gragg [1]. Thus, $c_F^{(1)}(T_{mn})y = d$ can be solved in $O(nm \log^2 m)$ operations.

Next we consider the cost of the matrix-vector multiplication $T_{mn}v$. We recall that for any Toeplitz matrix $T_{(k)} \in \mathbb{R}^{n \times n}$, the matrix-vector multiplication $T_{(k)}w$ can be computed by FFTs by first embedding $T_{(k)}w$ into a $2n \times 2n$ circulant matrix and extending w to a $2n$-vector by zeros, see Section 1.1.4. For the matrix-vector product $T_{mn}v$ we can use the same trick. We first embed T_{mn} into a (blockwise) $2m \times 2m$ block circulant matrix where each block itself is a $2n \times 2n$ circulant matrix. Then we extend v to a $4mn$-vector by putting zeros in the appropriate places. By using $F_{2m} \otimes F_{2n}$ to diagonalize the $4mn \times 4mn$ BCCB matrix, we see that $T_{mn}v$ can be obtained in $O(mn \log mn) = O(mn \log m + mn \log n)$ operations by using 2-dimensional FFTs. Thus we conclude that the initialization cost in this case is $O(mn \log n)$ and the cost per iteration is $O(nm \log^2 m + mn \log n)$. If $m > n$, we emphasize that one should consider using $\tilde{c}_F^{(1)}(A_{mn})$ as preconditioner instead.

Case of $c_{F,F}^{(2)}(T_{mn})$

For $c_{F,F}^{(2)}(T_{mn})$, by using (2.17), it is not difficult to find the k-th entry in the j-th block of the first column of the matrix is given by

$$c_k^{(j)} = \frac{1}{mn}[(m-j)(n-k)t_k^{(j)} + j(n-k)t_k^{(j-m)}$$

$$(2.24)$$

$$+ (m-j)kt_{k-n}^{(j)} + jkt_{k-n}^{(j-m)}],$$

for $0 \leq j \leq m-1$, $0 \leq k \leq n-1$. It requires only $O(mn)$ operations to compute the first column of $c_{F,F}^{(2)}(T_{mn})$. Followed by the same argument as in Section 2.2.1, we conclude that the initialization cost in this case is $O(mn \log mn)$ and the cost per iteration is still $O(mn \log mn)$.

2.3 Convergence rate

In this section, we analyse the convergence rate of the PCG method when applied to solving level-2 symmetric BTTB systems. Let us first consider general BTTB systems $T_{mn}u = b$. Let the entries of T_{mn} be denoted by

$$(T_{mn})_{p,q;r,s} = t_{p-q}^{(r-s)},$$

for $1 \leq p, q \leq n, 1 \leq r, s \leq m$. The T_{mn} is associated with a generating function $f(x, y)$ as follows,

$$t_k^{(j)}(f) \equiv \frac{1}{4\pi^2} \int_{-\pi}^{\pi} \int_{-\pi}^{\pi} f(x, y)e^{-i(jx+ky)} dx dy, \qquad i \equiv \sqrt{-1}.$$

We note that for any m and n, T_{mn}'s have the following important properties:

(i) When f is real-valued, then $T_{mn}(f)$ are Hermitian, i.e.,

$$t_k^{(j)}(f) = \bar{t}_{-k}^{(-j)}(f).$$

(ii) When f is real-valued with $f(x,y) = f(-x,-y)$, then $T_{mn}(f)$ are real symmetric, i.e.,

$$t_k^{(j)}(f) = t_{-k}^{(-j)}(f).$$

(iii) When f is real-valued and even, i.e., $f(x,y) = f(|x|,|y|)$, then $T_{mn}(f)$ are level-2 symmetric, i.e.,

$$t_k^{(j)}(f) = t_{|k|}^{(|j|)}(f).$$

Let $C_{2\pi \times 2\pi}$ denote the Banach space of all 2π-periodic (in each direction) continuous real-valued functions equipped with the supremum norm $||| \cdot |||_\infty$. The following theorem gives the relation between the values of $f(x,y)$ and the eigenvalues of $T_{mn}(f)$.

Theorem 2.4 *Let T_{mn} be a BTTB matrix with a generating function $f(x,y) \in C_{2\pi \times 2\pi}$. Then we have*

$$f_{\min} \leq \lambda_{\min}(T_{mn}) \leq \lambda_{\max}(T_{mn}) \leq f_{\max}$$

where $f_{\min}$ and $f_{\max}$ denote the minimum and maximum values of $f(x,y)$, respectively. In particular, if $f_{\min} > 0$ then T_{mn} is positive definite.

Proof: Let $u \in \mathbb{C}^{mn}$ be denoted by

$$u = (u_0^{(0)}, u_1^{(0)}, \cdots, u_{n-1}^{(0)}, u_0^{(1)}, \cdots, u_0^{(m-1)}, \cdots, u_{n-1}^{(m-1)})^T.$$

Then we have

$$u^* T_{mn} u = \frac{1}{4\pi^2} \int_{-\pi}^{\pi} \int_{-\pi}^{\pi} \left| \sum_{j=0}^{m-1} \sum_{k=0}^{n-1} u_k^{(j)} e^{-i(jx+ky)} \right|^2 f(x,y) dx dy. \qquad (2.25)$$

Since $f_{\min} \leq f(x,y) \leq f_{\max}$ for all x and y, we have by (2.25),

$$f_{\min} \leq u^* T_{mn} u \leq f_{\max}$$

provided that u is subjected to the condition

$$u^*u = \frac{1}{4\pi^2} \int_{-\pi}^{\pi} \int_{-\pi}^{\pi} \left| \sum_{j=0}^{m-1} \sum_{k=0}^{n-1} u_k^{(j)} e^{-i(jx+ky)} \right|^2 dx dy = 1.$$

Hence we have by Courant–Fischer's Minimax Theorem,

$$f_{\min} \leq \lambda_{\min}(T_{mn}) \leq \lambda_{\max}(T_{mn}) \leq f_{\max}. \quad \square$$

For T_{mn} given by (2.23), since T_{mn} is level-2 symmetric, the generating function $f(x,y)$ is assumed to be even. We also assume that $f(x,y)$ is in the Wiener class, i.e.,

$$\sum_{j=0}^{\infty} \sum_{k=0}^{\infty} |t_k^{(j)}(f)| < \infty.$$

We remark that the Wiener class of functions is a proper subset of $C_{2\pi \times 2\pi}$. Now, we are going to analyse the convergence rate of the PCG method when applied to solving level-2 symmetric BTTB systems.

2.3.1   Convergence rate of $c_F^{(1)}(T_{mn})$

In order to analyse the distribution of the eigenvalues of $T_{mn} - c_F^{(1)}(T_{mn})$ we need Strang's circulant preconditioner introduced in Section 1.2.1. For each $T_{(j)}$, by (1.7), the entries $s_{pq}^{(j)} = s_{|p-q|}^{(j)}$ of Strang's preconditioner $s(T_{(j)})$ are given by

$$s_k^{(j)} = \begin{cases} t_k^{(j)}, & 0 \leq k \leq r, \\ \\ t_{n-k}^{(j)}, & r < k < n. \end{cases} \tag{2.26}$$

Here for simplicity, we have assumed that $n = 2r + 1$. Define

$$s_F^{(1)}(T_{mn}) \equiv \begin{pmatrix} s(T_{(0)}) & s(T_{(1)}) & \cdots & s(T_{(m-1)}) \\ s(T_{(1)}) & s(T_{(0)}) & \cdots & s(T_{(m-2)}) \\ \vdots & \ddots & \ddots & \vdots \\ s(T_{(m-1)}) & s(T_{(m-2)}) & \cdots & s(T_{(0)}) \end{pmatrix}. \tag{2.27}$$

We prove below that the matrices $c_F^{(1)}(T_{mn})$ and $s_F^{(1)}(T_{mn})$ are asymptotically the same.

Lemma 2.4 *Let T_{mn} be given by (2.23) with a generating function f in the Wiener class. Then for all $m > 0$,*

$$\lim_{n \to \infty} \|s_F^{(1)}(T_{mn}) - c_F^{(1)}(T_{mn})\|_1 = 0.$$

Proof: Let

$$B_{mn} \equiv s_F^{(1)}(T_{mn}) - c_F^{(1)}(T_{mn}).$$

By (2.15) and (2.27), we see that the blocks $B_{(j)}$ of B_{mn} are given by

$$B_{(j)} \equiv s(T_{(j)}) - c_F(T_{(j)}),$$

for $j = 0, 1, \cdots, m-1$. Hence by (1.12) and (2.26), they are circulant with entries $b_{pq}^{(j)} = b_{|p-q|}^{(j)}$ given by

$$b_k^{(j)} = \begin{cases} \dfrac{k}{n}(t_k^{(j)} - t_{n-k}^{(j)}), & 0 \le k \le r, \\[2ex] \dfrac{n-k}{n}(t_{n-k}^{(j)} - t_k^{(j)}), & r < k < n. \end{cases}$$

Thus

$$\|B_{mn}\|_1 \le 2 \sum_{j=0}^{m-1} \|B_{(j)}\|_1 \le 2 \sum_{j=0}^{m-1} \sum_{k=0}^{n-1} |b_k^{(j)}|$$

$$\le 4 \sum_{j=0}^{m-1} \sum_{k=1}^{r} \frac{k}{n}|t_k^{(j)}| + 4 \sum_{j=0}^{m-1} \sum_{k=r+1}^{n-1} |t_k^{(j)}|.$$

For all $\varepsilon > 0$, since the generating function f is in the Wiener class, we can always find an $N_1 > 0$ and an $N_2 > 2N_1$, such that

$$\sum_{j=0}^{\infty} \sum_{k=N_1}^{\infty} |t_k^{(j)}(f)| < \varepsilon \quad \text{and} \quad \frac{1}{N_2} \sum_{j=0}^{\infty} \sum_{k=1}^{N_1} k|t_k^{(j)}(f)| < \varepsilon.$$

Thus for all $n > N_2$,

$$\|B_{mn}\|_1 \le \frac{4}{N_2} \sum_{j=0}^{\infty} \sum_{k=1}^{N_1} k|t_k^{(j)}| + 4 \sum_{j=0}^{\infty} \sum_{k=N_1+1}^{r} |t_k^{(j)}| + 4 \sum_{j=0}^{\infty} \sum_{k=r+1}^{\infty} |t_k^{(j)}|$$

$$< 12\varepsilon. \quad \square$$

In view of Lemma 2.4 and the following equality

$$T_{mn} - c_F^{(1)}(T_{mn}) = (s_F^{(1)}(T_{mn}) - c_F^{(1)}(T_{mn})) + (T_{mn} - s_F^{(1)}(T_{mn})),$$

we see that the spectra of $T_{mn} - c_F^{(1)}(T_{mn})$ and $T_{mn} - s_F^{(1)}(T_{mn})$ are asymptotically the same. However, it is easier to obtain spectral information about the matrix $T_{mn} - s_F^{(1)}(T_{mn})$ as the following lemma shows.

Lemma 2.5 *Let T_{mn} be given by (2.23) with a generating function f in the Wiener class. Then for all $\varepsilon > 0$, there exists an $N_3 > 0$, such that for all $n > N_3$ and for all $m > 0$,*

$$s_F^{(1)}(T_{mn}) - T_{mn} = W_{mn}^{(N_3)} + U_{mn}^{(N_3)}$$

where $\|W_{mn}^{(N_3)}\|_1 \le \varepsilon$ and $\mathrm{rank}(U_{mn}^{(N_3)}) \le O(m)$.

Proof: Let

$$W_{mn} \equiv s_F^{(1)}(T_{mn}) - T_{mn}.$$

It is clear from (2.26) that its blocks

$$W_{(j)} \equiv s(T_{(j)}) - T_{(j)},$$

for $j = 0, 1, \cdots, m-1$, are symmetric Toeplitz matrices with entries $w_{pq}^{(j)} = w_{|p-q|}^{(j)}$ given by

$$w_k^{(j)} = \begin{cases} 0, & 0 \le k \le r, \\ t_{n-k}^{(j)} - t_k^{(j)}, & r < k < n. \end{cases}$$

For all $\varepsilon > 0$, since the generating function f is in the Wiener class, there exists an $N_3 > 0$ such that

$$\sum_{j=0}^{\infty} \sum_{k=N_3}^{\infty} |t_k^{(j)}(f)| < \varepsilon.$$

Corresponding to this N_3, for each block $W_{(j)}$, we define the matrix

$$W_{(j)}^{(N_3)} = \begin{pmatrix} \tilde{W}_{(j)} & 0 \\ 0 & 0 \end{pmatrix} \in \mathbb{R}^{n \times n}$$

where $\tilde{W}_{(j)}$ is the $(n - N_3) \times (n - N_3)$ principal submatrix of $W_{(j)}$. Clearly, each $\tilde{W}_{(j)}$ is a Toeplitz matrix. Let

$$U_{(j)}^{(N_3)} \equiv W_{(j)} - W_{(j)}^{(N_3)}$$

for all j. We note that $U_{(j)}^{(N_3)}$ is non-zero only in the last N_3 rows and N_3 columns, therefore $\mathrm{rank}(U_{(j)}^{(N_3)}) \leq 2N_3$. Let

$$W_{mn}^{(N_3)} = \begin{pmatrix} W_{(0)}^{(N_3)} & W_{(1)}^{(N_3)} & \cdots & W_{(m-1)}^{(N_3)} \\ W_{(1)}^{(N_3)} & W_{(0)}^{(N_3)} & \cdots & W_{(m-2)}^{(N_3)} \\ \vdots & \ddots & \ddots & \vdots \\ W_{(m-1)}^{(N_3)} & W_{(m-2)}^{(N_3)} & \cdots & W_{(0)}^{(N_3)} \end{pmatrix} \tag{2.28}$$

and

$$U_{mn}^{(N_3)} = \begin{pmatrix} U_{(0)}^{(N_3)} & U_{(1)}^{(N_3)} & \cdots & U_{(m-1)}^{(N_3)} \\ U_{(1)}^{(N_3)} & U_{(0)}^{(N_3)} & \cdots & U_{(m-2)}^{(N_3)} \\ \vdots & \ddots & \ddots & \vdots \\ U_{(m-1)}^{(N_3)} & U_{(m-2)}^{(N_3)} & \cdots & U_{(0)}^{(N_3)} \end{pmatrix}. \tag{2.29}$$

Then

$$s_F^{(1)}(T_{mn}) - T_{mn} = W_{mn}^{(N_3)} + U_{mn}^{(N_3)}.$$

Since each block $U_{(j)}^{(N_3)} \in \mathbb{R}^{n \times n}$ in $U_{mn}^{(N_3)}$ is a matrix where the leading $(n - N_3) \times (n - N_3)$ principal submatrix is a zero matrix, it is easy to see that

$$\mathrm{rank}(U_{mn}^{(N_3)}) \leq 2N_3 m = O(m).$$

For $W_{mn}^{(N_3)}$, we have by (2.28),

$$\|W_{mn}^{(N_3)}\|_1 \leq 2 \sum_{j=0}^{m-1} \|W_{(j)}^{(N_3)}\|_1 = 2 \sum_{j=0}^{m-1} \|\tilde{W}_{(j)}\|_1$$

$$= 2 \sum_{j=0}^{m-1} \sum_{k=r+1}^{n-N_3-1} |w_k^{(j)}| = 2 \sum_{j=0}^{m-1} \sum_{k=r+1}^{n-N_3-1} |t_{n-k}^{(j)} - t_k^{(j)}|$$

$$\leq 2 \sum_{j=0}^{m-1} \sum_{k=N_3+1}^{n-N_3-1} |t_k^{(j)}| \leq 2 \sum_{j=0}^{\infty} \sum_{k=N_3}^{\infty} |t_k^{(j)}|$$

$$< 2\varepsilon. \quad \square$$

Let $N = \max(N_2, N_3)$ where N_2 and N_3 are given in the proofs of Lemmas 2.4 and 2.5. Then for all $n > N$ and $m > 0$, we have

$$T_{mn} - c_F^{(1)}(T_{mn}) = M_{mn} + L_{O(m)}, \tag{2.30}$$

where

$$M_{mn} \equiv s_F^{(1)}(T_{mn}) - c_F^{(1)}(T_{mn}) + W_{mn}^{(N)}$$

with $\|M_{mn}\|_1 < \varepsilon$, and

$$L_{O(m)} \equiv U_{mn}^{(N)}$$

with $\text{rank}(L_{O(m)}) \leq O(m)$. Since M_{mn} is symmetric, we have

$$\|M_{mn}\|_2 \leq (\|M_{mn}\|_1 \|M_{mn}\|_\infty)^{\frac{1}{2}} = \|M_{mn}\|_1 < \varepsilon.$$

By using Weyl's Theorem we then have the following theorem.

Theorem 2.5 *Let T_{mn} be given by (2.23) with a generating function f in the Wiener class. Then for all $\varepsilon > 0$ there exists an $N > 0$ such that for all $n > N$ and all $m > 0$, at most $O(m)$ eigenvalues of $c_F^{(1)}(T_{mn}) - T_{mn}$ have absolute values larger than ε.*

Assume that $f_{\min} > 0$, by Theorem 2.4, T_{mn} is positive definite with

$$f_{\min} \leq \lambda_{\min}(T_{mn}) \leq \lambda_{\max}(T_{mn}) \leq f_{\max}.$$

By Theorem 2.1 (iv) we then have

$$0 < f_{\min} \leq \lambda_{\min}(c_F^{(1)}(T_{mn})) \leq \lambda_{\max}(c_F^{(1)}(T_{mn})) \leq f_{\max}.$$

Hence $\|(c_F^{(1)}(T_{mn}))^{-1}\|_2$ is uniformly bounded. By noting that

$$(c_F^{(1)}(T_{mn}))^{-1}T_{mn} = I - (c_F^{(1)}(T_{mn}))^{-1}(c_F^{(1)}(T_{mn}) - T_{mn}),$$

we have the following immediate corollary.

Corollary 2.1 *Let T_{mn} be given by (2.23) with a generating function $f > 0$ in the Wiener class. Then for all $\varepsilon > 0$, there exists an $N > 0$ such that for all $n > N$ and all $m > 0$, at most $O(m)$ eigenvalues of $(c_F^{(1)}(T_{mn}))^{-1}T_{mn} - I$ have the absolute value larger than ε.*

As a consequence the spectrum of $(c_F^{(1)}(T_{mn}))^{-1}T_{mn}$ is clustered around 1 except for at most $O(m)$ outlying eigenvalues which are also bounded. When the PCG method is applied to solving $T_{mn}u = b$, by Theorem 1.9 and Corollary 2.1 we know that the number of iterations is independent of m and n, and the convergence rate will be fast. We recall that in Section 2.2.2 the algorithm requires $O(mn \log n)$ operations in the initialization step and $O(mn \log^2 m + mn \log n)$ operations in each iteration. Thus the total complexity of the algorithm remains $O(mn \log^2 m + mn \log n)$. Our theoretical results are confirmed by numerical tests in Section 2.5.

2.3.2 Convergence rate of $c_{F,F}^{(2)}(T_{mn})$

For the convergence rate of the PCG method with $c_{F,F}^{(2)}(T_{mn})$, we note that

$$T_{mn} - c_{F,F}^{(2)}(T_{mn}) = T_{mn} - \tilde{c}_F^{(1)}(T_{mn}) + \tilde{c}_F^{(1)}(T_{mn}) - c_{F,F}^{(2)}(T_{mn})$$

$$= T_{mn} - \tilde{c}_F^{(1)}(T_{mn}) + \tilde{c}_F^{(1)}(T_{mn}) - (\tilde{c}_F^{(1)} \circ c_F^{(1)})(T_{mn})$$

$$= T_{mn} - \tilde{c}_F^{(1)}(T_{mn}) + \tilde{c}_F^{(1)}\left(T_{mn} - c_F^{(1)}(T_{mn})\right)$$

$$= M_{mn} + L_{O(n)} + \tilde{c}_F^{(1)}(N_{mn} + L_{O(m)})$$

where M_{mn}, N_{mn}, $L_{O(m)}$ and $L_{O(n)}$ are defined similarly as in (2.30) with

$$\|M_{mn}\|_1 < \varepsilon, \quad \|N_{mn}\|_1 < \varepsilon,$$

and

$$\mathrm{rank}(L_{O(m)}) \le O(m), \quad \mathrm{rank}(L_{O(n)}) \le O(n).$$

Since

$$\tilde{c}_F^{(1)}(N_{mn} + L_{O(m)}) = \tilde{c}_F^{(1)}(N_{mn}) + \tilde{c}_F^{(1)}(L_{O(m)}), \qquad (2.31)$$

for the first term in the right hand side of (2.31), we have

$$\|\tilde{c}_F^{(1)}(N_{mn})\|_2 \le \|\tilde{c}_F^{(1)}\|_2 \|N_{mn}\|_2 \le \|N_{mn}\|_2 \le \|N_{mn}\|_1 < \varepsilon.$$

For the second term in the right hand side of (2.31), by noting (2.16) and (2.29) one can easily show that

$$\mathrm{rank}\left(\tilde{c}_F^{(1)}(L_{O(m)})\right) \le O(m).$$

We therefore have the following theorem.

Theorem 2.6 *Let T_{mn} be given by (2.23) with a generating function f in the Wiener class. Then for all $\varepsilon > 0$ there exist $M, N > 0$ such that for all $m > M$ and all $n > N$, at most $O(m) + O(n)$ eigenvalues of $c_{F,F}^{(2)}(T_{mn}) - T_{mn}$ have absolute values larger than ε.*

When f is positive, by Theorem 2.4 and Theorem 2.3 (ii), we have

$$0 < f_{\min} \le \lambda_{\min}(c_{F,F}^{(2)}(T_{mn})) \le \lambda_{\max}(c_{F,F}^{(2)}(T_{mn})) \le f_{\max}.$$

Hence $\|(c_{F,F}^{(2)}(T_{mn}))^{-1}\|_2$ is uniformly bounded. By noting that

$$(c_{F,F}^{(2)}(T_{mn}))^{-1}T_{mn} = I - (c_{F,F}^{(2)}(T_{mn}))^{-1}(c_{F,F}^{(2)}(T_{mn}) - T_{mn}),$$

we then have following immediate corollary.

Corollary 2.2 *Let T_{mn} be given by (2.23) with a generating function $f > 0$ in the Wiener class. Then for all $\varepsilon > 0$, there exist $M, N > 0$, such that for all $m > M$ and all $n > N$, at most $O(m) + O(n)$ eigenvalues of $(c_{F,F}^{(2)}(T_{mn}))^{-1}T_{mn} - I$ have absolute value larger than ε.*

As a consequence the spectrum of $(c^{(2)}_{F,F}(T_{mn}))^{-1}T_{mn}$ is clustered around 1 except for at most $O(m) + O(n)$ outlying eigenvalues which are also bounded. When the PCG method is applied to solving $T_{mn}u = b$, by Theorem 1.9 and Corollary 2.2 we know that the number of iterations is independent of m and n, and the convergence rate will be fast. We recall that in Section 2.2.2 the algorithm requires $O(mn \log mn)$ operations in both initialization step and each iteration step. Thus the total complexity of the algorithm remains $O(mn \log mn)$.

Finally, we would like to point out that for level-2 symmetric block Toeplitz matrices T_{mn}, we can define, analogously to $\tilde{c}^{(1)}_V(T_{mn})$, the matrix $\tilde{s}^{(1)}_F(T_{mn})$ as follows,

$$\tilde{s}^{(1)}_F(T_{mn}) \equiv P^* s^{(1)}_F(PT_{mn}P^*)P$$

where P is given by (2.6). Then as in Section 2.1.3 we can further define the BCCB preconditioner

$$s^{(2)}_{F,F}(T_{mn}) \equiv \tilde{s}^{(1)}_F \circ s^{(1)}_F(T_{mn}) \tag{2.32}$$

which is an approximation of T_{mn} in two directions as remarked after the proof of Theorem 2.3. If, instead of Strang's circulant precondi- tioner, R. Chan's preconditioner defined as in Section 1.2.3 is used in (2.27) and (2.32), we then have three preconditioners $r^{(1)}_F(T_{mn})$, $\tilde{r}^{(1)}_F(T_{mn})$ and $r^{(2)}_{F,F}(T_{mn})$, see Section 2.4 for a detail. The BCCB preconditioner $r^{(2)}_{F,F}(T_{mn})$ is the preconditioner considered in [65]. We would like to men- tion that in general, BCCB preconditioners for BTTB systems are not optimal in the sense that spectra of preconditioned matrices are not clus- tered around 1, see [83].

2.4   Invertibility of $r^{(2)}_{F,F}(T_{mn})$ and $s^{(2)}_{F,F}(T_{mn})$

Our task in this section is to show that for T_{mn} given by (2.23), the BCCB preconditioners $r^{(2)}_{F,F}(T_{mn})$ and $s^{(2)}_{F,F}(T_{mn})$ are uniformly invertible when m and n are sufficiently large. Moreover, we will prove that the spectra of the BCCB preconditioners $c^{(2)}_{F,F}(T_{mn})$, $r^{(2)}_{F,F}(T_{mn})$ and $s^{(2)}_{F,F}(T_{mn})$ are asymptotically the same. Thus, when the PCG method is applied, all of these preconditioned systems converge at the same rate for large m and n.

We first discuss the preconditioner $r_{F,F}^{(2)}(T_{mn})$. For each $T_{(j)}$, $j = 0, \cdots, m-1$, let $r_F(T_{(j)})$ be the circulant matrix proposed as in (1.13) with diagonals $r_k^{(j)} = r_{-k}^{(j)}$ given by

$$r_k^{(j)} = t_{k-n}^{(j)} + t_k^{(j)} \tag{2.33}$$

for $0 \le k < n$, where $t_{-n}^{(j)}$ is taken to be 0. Two operators $r_F^{(1)}$ and $\tilde{r}_F^{(1)}$ are defined as follows,

$$r_F^{(1)}(T_{mn}) \equiv \begin{pmatrix} r_F(T_{(0)}) & r_F(T_{(1)}) & \cdots & r_F(T_{(m-1)}) \\ r_F(T_{(1)}) & r_F(T_{(0)}) & \cdots & r_F(T_{(m-2)}) \\ \vdots & \ddots & \ddots & \vdots \\ r_F(T_{(m-1)}) & r_F(T_{(m-2)}) & \cdots & r_F(T_{(0)}) \end{pmatrix} \tag{2.34}$$

and

$$\tilde{r}_F^{(1)}(T_{mn}) \equiv P^* r_F^{(1)}(P T_{mn} P^*) P$$

where P is given by (2.6). Let

$$r_{F,F}^{(2)}(T_{mn}) \equiv \tilde{r}_F^{(1)} \circ r_F^{(1)}(T_{mn}). \tag{2.35}$$

For the relation between $c_F^{(1)}$ and $r_F^{(1)}$, and the relation between $\tilde{c}_F^{(1)}$ and $\tilde{r}_F^{(1)}$, we have

Lemma 2.6 *Let T_{mn} be given by (2.23) with a generating function f in the Wiener class. Then for all $m > 0$, we have*

$$\lim_{n\to\infty} \|c_F^{(1)}(T_{mn}) - r_F^{(1)}(T_{mn})\|_2 = 0. \tag{2.36}$$

For all $n > 0$, we have

$$\lim_{m\to\infty} \|\tilde{c}_F^{(1)}(T_{mn}) - \tilde{r}_F^{(1)}(T_{mn})\|_2 = 0. \tag{2.37}$$

Proof: We only prove (2.36) and the proof of (2.37) is similar. Let

$$B_{mn} \equiv r_F^{(1)}(T_{mn}) - c_F^{(1)}(T_{mn}).$$

By (2.15) and (2.34), we know that each block $B_{(j)}$ of B_{mn} is given by

$$B_{(j)} \equiv r_F(T_{(j)}) - c_F(T_{(j)}),$$

for $j = 0, \cdots, m - 1$. Hence by (2.33) and (1.12), each $B_{(j)}$ is circulant with entries $b_{pq}^{(j)} = b_{|p-q|}^{(j)}$ given by

$$b_k^{(j)} = \frac{(n-k)t_{n-k}^{(j)} + k t_k^{(j)}}{n}.$$

Thus,

$$\|B_{mn}\|_1 \leq 2 \sum_{j=0}^{m-1} \|B_{(j)}\|_1 \leq 2 \sum_{j=0}^{m-1} \sum_{k=0}^{n-1} |b_k^{(j)}| \leq 4 \sum_{j=0}^{m-1} \sum_{k=1}^{n-1} \frac{k}{n} |t_k^{(j)}|.$$

For all $\varepsilon > 0$, since the generating function f is in the Wiener class, i.e.,

$$\sum_{j=0}^{\infty} \sum_{k=0}^{\infty} |t_k^{(j)}(f)| < \infty,$$

we can always find an $N_1 > 0$ and an $N_2 > 2N_1$ such that

$$\sum_{j=0}^{\infty} \sum_{k=N_1+1}^{\infty} |t_k^{(j)}| < \frac{\varepsilon}{8} \quad \text{and} \quad \frac{1}{N_2} \sum_{j=0}^{\infty} \sum_{k=1}^{N_1} k|t_k^{(j)}| < \frac{\varepsilon}{8}.$$

Thus for all $n > N_2$,

$$\|B_{mn}\|_1 \leq \frac{4}{N_2} \sum_{j=0}^{\infty} \sum_{k=1}^{N_1} k|t_k^{(j)}| + 4 \sum_{j=0}^{\infty} \sum_{k=N_1+1}^{\infty} |t_k^{(j)}| < \varepsilon.$$

Since B_{mn} is symmetric we have

$$\|B_{mn}\|_2 \leq (\|B_{mn}\|_1 \|B_{mn}\|_\infty)^{1/2} = \|B_{mn}\|_1 \leq \varepsilon. \quad \square$$

We prove below that the spectra of $c_{F,F}^{(2)}(T_{mn})$ and $r_{F,F}^{(2)}(T_{mn})$ are asymptotically the same.

Theorem 2.7 *Let T_{mn} be given by (2.23) with a generating function f in the Wiener class. Then for all $\varepsilon > 0$, when m and n are sufficiently large we have*

$$\|c_{F,F}^{(2)}(T_{mn}) - r_{F,F}^{(2)}(T_{mn})\|_2 \leq \varepsilon.$$

Proof: Since by Theorem 2.2 (iv), $\|\tilde{c}_F^{(1)}\|_2 = 1$, we then have

$$\|c_{F,F}^{(2)}(T_{mn}) - r_{F,F}^{(2)}(T_{mn})\|_2$$

$$= \|(\tilde{c}_F^{(1)} \circ c_F^{(1)} - \tilde{c}_F^{(1)} \circ r_F^{(1)} + \tilde{c}_F^{(1)} \circ r_F^{(1)} - \tilde{r}_F^{(1)} \circ r_F^{(1)})(T_{mn})\|_2$$

$$\leq \|\tilde{c}_F^{(1)}((c_F^{(1)} - r_F^{(1)})(T_{mn}))\|_2$$

$$\quad + \|(\tilde{c}_F^{(1)} \circ r_F^{(1)} - \tilde{r}_F^{(1)} \circ r_F^{(1)})(T_{mn})\|_2$$

$$\leq \|\tilde{c}_F^{(1)}\|_2 \|c_F^{(1)}(T_{mn}) - r_F^{(1)}(T_{mn})\|_2$$

$$\quad + \|\tilde{c}_F^{(1)}(r_F^{(1)}(T_{mn})) - \tilde{r}_F^{(1)}(r_F^{(1)}(T_{mn}))\|_2$$

$$= \|c_F^{(1)}(T_{mn}) - r_F^{(1)}(T_{mn})\|_2$$

$$\quad + \|\tilde{c}_F^{(1)}(r_F^{(1)}(T_{mn})) - \tilde{r}_F^{(1)}(r_F^{(1)}(T_{mn}))\|_2. \tag{2.38}$$

By (2.36) we have

$$\|c_F^{(1)}(T_{mn}) - r_F^{(1)}(T_{mn})\|_2 \leq \frac{\varepsilon}{2} \tag{2.39}$$

for n sufficiently large. For the last line of (2.38), when n is fixed, $r_F^{(1)}(T_{mn})$ is a level-2 symmetric BTTB matrix with an absolutely summable generating sequence which satisfies

$$\sum_{j=0}^{\infty} \sum_{k=0}^{n} |r_k^{(j)}| \leq 2 \sum_{j=0}^{\infty} \sum_{k=0}^{\infty} |t_k^{(j)}(f)| < \infty.$$

By (2.37) we have

$$\|\tilde{c}_F^{(1)}(r_F^{(1)}(T_{mn})) - \tilde{r}_F^{(1)}(r_F^{(1)}(T_{mn}))\|_2 \leq \frac{\varepsilon}{2} \tag{2.40}$$

for m sufficiently large. Combining (2.39) and (2.40) we have our result.

□

By using Theorem 2.4, Theorem 2.3 (ii), and Theorem 2.7 we then have

Corollary 2.3 *Let T_{mn} be given by (2.23) with a generating function $f > 0$ in the Wiener class. Then the preconditioner $r_{F,F}^{(2)}(T_{mn})$ is positive definite when m and n are sufficiently large.*

Finally, we briefly discuss the invertibility of the preconditioner $s_{F,F}^{(2)}(T_{mn})$. By Lemma 2.4 we have already known that the matrices $c_F^{(1)}(T_{mn})$ and $s_F^{(1)}(T_{mn})$ are asymptotically the same. By using the same trick as in the proof of Theorem 2.7 one can prove the following theorem easily.

Theorem 2.8 *Let T_{mn} be given by (2.23) with a generating function f in the Wiener class. Then for all $\varepsilon > 0$, when m and n are sufficiently large we have*

$$\|c_{F,F}^{(2)}(T_{mn}) - s_{F,F}^{(2)}(T_{mn})\|_2 \le \varepsilon.$$

From Theorem 2.8, we know that the spectra of $c_{F,F}^{(2)}(T_{mn})$ and $s_{F,F}^{(2)}(T_{mn})$ are asymptotically the same. We then have

Corollary 2.4 *Let T_{mn} be given by (2.23) with a generating function $f > 0$ in the Wiener class. Then the preconditioner $s_{F,F}^{(2)}(T_{mn})$ is positive definite when m and n are sufficiently large.*

Since the spectra of the preconditioners $c_{F,F}^{(2)}(T_{mn})$, $r_{F,F}^{(2)}(T_{mn})$ and $s_{F,F}^{(2)}(T_{mn})$ are asymptotically the same, we know that all of these preconditioned systems converge at the same rate for large m and n.

2.5   Numerical results

In this section, we apply the PCG method to level-2 symmetric BTTB systems $T_{mn}u = b$ where T_{mn} are of the form as in (2.23) with the diagonals of the blocks $T_{(j)}$ being given by $t_i^{(j)}$. Four different generating sequences were tested. They were:

(i) $t_i^{(j)} = \dfrac{1}{(j+1)(|i|+1)^{1+0.1\times(j+1)}}$, $j \geq 0,\ i = 0, \pm 1, \pm 2, \cdots$;

(ii) $t_i^{(j)} = \dfrac{1}{(j+1)^{1.1}(|i|+1)^{1+0.1\times(j+1)}}$, $j \geq 0,\ i = 0, \pm 1, \pm 2, \cdots$;

(iii) $t_i^{(j)} = \dfrac{1}{(j+1)^{1.1} + (|i|+1)^{1.1}}$, $j \geq 0,\ i = 0, \pm 1, \pm 2, \cdots$;

(iv) $t_i^{(j)} = \dfrac{1}{(j+1)^{2.1} + (|i|+1)^{2.1}}$, $j \geq 0,\ i = 0, \pm 1, \pm 2, \cdots$.

The generating sequences (ii) and (iv) are absolutely summable whilst (i) and (iii) are not. Tables 2.1 and 2.2 show the number of iterations required for convergence. In these tables, the notation I means that no preconditioner is used. The stopping criteria for the method is set at

$$\frac{\|r^{(k)}\|_2}{\|r^{(0)}\|_2} < 10^{-7}$$

where $r^{(k)}$ is the residual vector at the k-th iteration. The right hand side vector b is chosen to be the vector of all 1's and the zero vector is the initial guess. In all cases, we see that as $m = n$ increases, the number of iterations remains roughly a constant for both $c_F^{(1)}(T_{mn})$ and $c_{F,F}^{(2)}(T_{mn})$.

Table 2.1. Preconditioners used and number of iterations

$n = m$	mn	Sequence (i)			Sequence (ii)		
		I	$c_F^{(1)}(T_{mn})$	$c_{F,F}^{(2)}(T_{mn})$	I	$c_F^{(1)}(T_{mn})$	$c_{F,F}^{(2)}(T_{mn})$
8	64	15	6	7	15	5	7
16	256	28	6	8	27	6	8
32	1024	37	6	8	35	6	8
64	4096	45	7	9	42	7	9
128	16384	50	7	9	46	7	9

Table 2.2. Preconditioners used and number of iterations

$n = m$	mn	Sequence (iii)			Sequence (iv)		
		I	$c_F^{(1)}(T_{mn})$	$c_{F,F}^{(2)}(T_{mn})$	I	$c_F^{(1)}(T_{mn})$	$c_{F,F}^{(2)}(T_{mn})$
8	64	11	7	7	10	7	7
16	256	27	8	8	16	8	7
32	1024	42	9	8	23	9	8
64	4096	71	11	9	30	9	8
128	16384	101	12	9	37	8	8

Chapter 3

BCCB Preconditioners from Kernels

In this chapter, a unified treatment for constructing BCCB preconditioners from viewpoint of kernels is given. For solving BTTB systems, we show that most of the well known BCCB preconditioners can be derived from convoluting the generating functions of systems with some famous kernels. A convergence analysis is also given. The initial idea of this unifying approach comes from [29].

3.1 Introduction

As in Section 2.3, let $\mathcal{C}_{2\pi \times 2\pi}$ denote the Banach space of all 2π-periodic continuous real-valued functions $f(x, y)$ equipped with the supremum norm $||| \cdot |||_\infty$. We first extend the result in Theorem 2.6 from the Wiener class to $\mathcal{C}_{2\pi \times 2\pi}$.

Theorem 3.1 *Let T_{mn} be a BTTB matrix with a generating function $f \in \mathcal{C}_{2\pi \times 2\pi}$. Then for all $\epsilon > 0$, there exist M and $N > 0$ such that for all $m > M$ and $n > N$, at most $O(m) + O(n)$ eigenvalues of $T_{mn}(f) - c_{F,F}^{(2)}(T_{mn})(f)$ have absolute values larger than ϵ.*

Proof:. Since $f \in \mathcal{C}_{2\pi \times 2\pi}$, by Stone–Weierstrass Theorem, see [72], we note that for any $\epsilon > 0$, there is a trigonometric polynomial

$$p_{MN}(x,y) = \sum_{j=-M}^{M} \sum_{k=-N}^{N} b_{j,k} e^{i(jx+ky)}$$

with $b_{-j,-k} = \bar{b}_{j,k}$ for $\forall |j| \leq M$ and $\forall |k| \leq N$, such that

$$|||f - p_{MN}|||_\infty \leq \epsilon. \tag{3.1}$$

For all $m > 2M$ and $n > 2N$, we have

$$c_{F,F}^{(2)}(T_{mn})(f) - T_{mn}(f) = c_{F,F}^{(2)}(T_{mn})(f - p_{MN}) - T_{mn}(f - p_{MN})$$
$$+ c_{F,F}^{(2)}(T_{mn})(p_{MN}) - T_{mn}(p_{MN}). \tag{3.2}$$

For the first two terms in the right hand side of (3.2), we note that by Theorem 2.3 (ii), Theorem 2.4 and (3.1),

$$\|c_{F,F}^{(2)}(T_{mn})(f - p_{MN}) - T_{mn}(f - p_{MN})\|_2$$

$$\leq \|c_{F,F}^{(2)}(T_{mn})(f - p_{MN})\|_2 + \|T_{mn}(f - p_{MN})\|_2$$

$$\leq |||f - p_{MN}|||_\infty + |||f - p_{MN}|||_\infty$$

$$\leq 2\epsilon.$$

Since p_{MN} is a real-valued function in the Wiener class, by Theorem 2.6 the matrix

$$c_{F,F}^{(2)}(T_{mn})(p_{MN}) - T_{mn}(p_{MN})$$

has a clustered spectrum. Hence by using Weyl's Theorem, the result follows. $\square$

By noting that

$$(c_{F,F}^{(2)}(T_{mn}))^{-1}T_{mn} - I_{mn} = (c_{F,F}^{(2)}(T_{mn}))^{-1}(T_{mn} - c_{F,F}^{(2)}(T_{mn})),$$

we have the following corollary.

Corollary 3.1 *Let T_{mn} be a BTTB matrix with a generating function $f > 0$ in $C_{2\pi \times 2\pi}$. Then for all $\epsilon > 0$ there exist M and $N > 0$ such that for all $m > M$ and $n > N$, at most $O(m) + O(n)$ eigenvalues of the matrix $(c_{F,F}^{(2)}(T_{mn}))^{-1}T_{mn} - I_{mn}$ have absolute values larger than ϵ.*

The BCCB preconditioners studied in last chapter have been shown that they are good preconditioners for solving a large class of BTTB systems by the PCG method. In the next section we try to construct the BCCB preconditioners from viewpoint of kernels and then discuss the convergence rate of the PCG method when it is applied to solving the system $T_{mn}(f)u = b$.

3.2   Preconditioners from kernels

Let $T_{mn}(f)$ be the BTTB matrix whose diagonals are given by

$$t_{j,k}(f) \equiv t_k^{(j)}(f) = \frac{1}{4\pi^2} \int_{-\pi}^{\pi} \int_{-\pi}^{\pi} f(x,y)e^{-i(jx+ky)}\,dxdy,$$

for $j,k = 0, \pm 1, \pm 2, \cdots$, and $f \in \mathcal{C}_{2\pi \times 2\pi}$.

3.2.1 $s_{F,F}^{(2)}(T_{mn})$

For Strang's BCCB preconditioner $s_{F,F}^{(2)}(T_{mn})(f)$ defined as in (2.32), the first column of it is given by (for simplicity, let $m = 2M + 1$ and $n = 2N + 1$),

$$c_{j,k} = \begin{cases} t_{j,k}, & 0 \le j \le M, \ 0 \le k \le N, \\[2mm] t_{j-m,k}, & M < j < m, 0 \le k \le N, \\[2mm] t_{j,k-n}, & 0 \le j \le M, \ N < k < n, \\[2mm] t_{j-m,k-n}, & M < j < m, N < k < n, \end{cases} \tag{3.3}$$

see (1.7) and (2.32). By (2.22) and (3.3), the eigenvalues of $s_{F,F}^{(2)}(T_{mn})(f)$ are given as follows,

$$\lambda_{p,q}\left(s_{F,F}^{(2)}(T_{mn})(f)\right) = \sum_{j=0}^{m-1}\sum_{k=0}^{n-1} c_{j,k}\xi_p^j\eta_q^k$$

$$= \sum_{j=0}^{M}\sum_{k=0}^{N} t_{j,k}\xi_p^j\eta_q^k + \sum_{j=M+1}^{m-1}\sum_{k=0}^{N} t_{j-m,k}\xi_p^j\eta_q^k$$

$$+ \sum_{j=0}^{M}\sum_{k=N+1}^{n-1} t_{j,k-n}\xi_p^j\eta_q^k + \sum_{j=M+1}^{m-1}\sum_{k=N+1}^{n} t_{j-m,k-n}\xi_p^j\eta_q^k$$

$$= \sum_{j=0}^{M}\sum_{k=0}^{N} t_{j,k}\xi_p^j\eta_q^k + \sum_{j=1}^{M}\sum_{k=0}^{N} t_{-j,k}\xi_p^{-j}\eta_q^k$$

$$+ \sum_{j=0}^{M}\sum_{k=1}^{N} t_{j,-k}\xi_p^j\eta_q^{-k} + \sum_{j=1}^{M}\sum_{k=1}^{N} t_{-j,-k}\xi_p^{-j}\eta_q^{-k}$$

$$= \sum_{j=-M}^{M}\sum_{k=-N}^{N} t_{j,k}\xi_p^j\eta_q^k = s_{MN}(f)\left(\frac{2\pi p}{m},\frac{2\pi q}{n}\right)$$

where

$$s_{MN}(f)(x,y) \equiv \sum_{j=-M}^{M}\sum_{k=-N}^{N} t_{j,k}e^{i(jx+ky)}. \tag{3.4}$$

Since

$$s_{MN}(f)(x,y)$$
$$= \frac{1}{4\pi^2}\int_{-\pi}^{\pi}\int_{-\pi}^{\pi} f(s,t)\frac{\sin\frac{(2M+1)(x-s)}{2}}{\sin\frac{x-s}{2}}\frac{\sin\frac{(2N+1)(y-t)}{2}}{\sin\frac{y-t}{2}}dsdt$$

$$= \frac{1}{4\pi^2}\int_{-\pi}^{\pi}\int_{-\pi}^{\pi} f(s,t)\widehat{D_M}(x-s)\widehat{D_N}(y-t)dsdt$$

where $\widehat{D_M}$ is the Dirichlet kernel, see [101], we then have

$$\lambda_{p,q}\left(s_{F,F}^{(2)}(T_{mn})(f)\right)$$

$$= \frac{1}{4\pi^2}\int_{-\pi}^{\pi}\int_{-\pi}^{\pi} f(s,t)\widehat{D_M}\left(\frac{2\pi p}{m}-s\right)\widehat{D_N}\left(\frac{2\pi q}{n}-t\right)dsdt$$

$$= f * \widehat{D_{MN}}\left(\frac{2\pi p}{m},\frac{2\pi q}{n}\right).$$

Here '*' denotes the convolution and

$$\widehat{D_{MN}}(x,y) \equiv \widehat{D_M}(x)\cdot\widehat{D_N}(y). \tag{3.5}$$

3.2.2 $c_{F,F}^{(2)}(T_{mn})$

For T. Chan's BCCB preconditioner $c_{F,F}^{(2)}(T_{mn})(f)$ defined as in (2.17), the first column of it is given by

$$
c_{j,k} = \frac{1}{mn}[(m-j)(n-k)t_{j,k} + j(n-k)t_{j-m,k}
$$

$$
+(m-j)kt_{j,k-n} + jkt_{j-m,k-n}],
\tag{3.6}
$$

for $0 \le j \le m-1$, $0 \le k \le n-1$, see (2.24). By (2.22) and (3.6) the eigenvalues of $c_{F,F}^{(2)}(T_{mn})(f)$ are given as follows,

$$
\lambda_{p,q}\left(c_{F,F}^{(2)}(T_{mn})(f)\right) = \sum_{j=0}^{m-1}\sum_{k=0}^{n-1} c_{j,k}\xi_p^j\eta_q^k
$$

$$
= \sum_{j=0}^{m-1}\sum_{k=0}^{n-1} \frac{(m-j)(n-k)}{mn} t_{j,k}\xi_p^j\eta_q^k
$$

$$
+ \sum_{j=1}^{m-1}\sum_{k=0}^{n-1} \frac{j(n-k)}{mn} t_{j-m,k}\xi_p^j\eta_q^k
$$

$$
+ \sum_{j=0}^{m-1}\sum_{k=1}^{n-1} \frac{(m-j)k}{mn} t_{j,k-n}\xi_p^j\eta_q^k
$$

$$
+ \sum_{j=1}^{m-1}\sum_{k=1}^{n-1} \frac{jk}{mn} t_{j-m,k-n}\xi_p^j\eta_q^k
$$

$$
= \sum_{j=0}^{m-1}\sum_{k=0}^{n-1} \frac{(m-j)(n-k)}{mn} t_{j,k}\xi_p^j\eta_q^k
$$

$$
+ \sum_{j=1}^{m-1}\sum_{k=0}^{n-1} \frac{(m-j)(n-k)}{mn} t_{-j,k}\xi_p^{-j}\eta_q^k
$$

$$+ \sum_{j=0}^{m-1} \sum_{k=1}^{n-1} \frac{(m-j)(n-k)}{mn} t_{j,-k} \xi_p^j \eta_q^{-k}$$

$$+ \sum_{j=1}^{m-1} \sum_{k=1}^{n-1} \frac{(m-j)(n-k)}{mn} t_{-j,-k} \xi_p^{-j} \eta_q^{-k}$$

$$= \frac{1}{mn} \sum_{j=-(m-1)}^{m-1} \sum_{k=-(n-1)}^{(n-1)} (m-|j|)(n-|k|) t_{j,k} \xi_p^j \eta_q^k$$

$$= \frac{1}{mn} \sum_{j=1}^{m-1} \sum_{k=1}^{n-1} s_{jk}(f)(2\pi p/m, 2\pi q/n)$$

where $s_{jk}(f)(x, y)$ is the partial sum defined as in (3.4). Since

$$\frac{1}{mn} \sum_{j=1}^{m-1} \sum_{k=1}^{n-1} s_{jk}(f)(x, y)$$

$$= \frac{1}{4\pi^2} \int_{-\pi}^{\pi} \int_{-\pi}^{\pi} f(s,t) \frac{1}{m} \left[\frac{\sin \frac{m(x-s)}{2}}{\sin \frac{x-s}{2}} \right]^2 \frac{1}{n} \left[\frac{\sin \frac{n(y-t)}{2}}{\sin \frac{y-t}{2}} \right]^2 ds\,dt$$

$$= \frac{1}{4\pi^2} \int_{-\pi}^{\pi} \int_{-\pi}^{\pi} f(s,t) \widehat{F_m}(x-s) \widehat{F_n}(y-t) ds\,dt$$

where $\widehat{F_m}$ is the Fejer kernel, see [101], we then have

$$\lambda_{p,q} \left(c_{F,F}^{(2)}(T_{mn})(f) \right)$$

$$= \frac{1}{4\pi^2} \int_{-\pi}^{\pi} \int_{-\pi}^{\pi} f(s,t) \widehat{F_m} \left(\frac{2\pi p}{m} - s \right) \widehat{F_n} \left(\frac{2\pi q}{n} - t \right) ds\,dt$$

$$= f * \widehat{F_{mn}}(2\pi p/m, 2\pi q/n)$$

where

$$\widehat{F_{mn}}(x, y) \equiv \widehat{F_m}(x) \cdot \widehat{F_n}(y). \tag{3.7}$$

3.2.3 $r_{F,F}^{(2)}(T_{mn})$

For R. Chan's BCCB preconditioner $r_{F,F}^{(2)}(T_{mn})(f)$ defined as in (2.35), the first column of it is given by

$$
\begin{cases}
c_{0,0} = t_{0,0}, & \\
c_{0,k} = t_{0,k} + t_{0,k-n}, & 1 \le k \le n-1, \\
c_{j,0} = t_{j,0} + t_{j-m,0}, & 1 \le j \le m-1, \\
c_{j,k} = t_{j,k} + t_{j-m,k} + t_{j,k-n} + t_{j-m,k-n}, & 1 \le j \le m-1, \\
& 1 \le k \le n-1,
\end{cases}
\tag{3.8}
$$

see (1.13) and (2.35).

By (2.22) and (3.8) the eigenvalues of

$$
r_{F,F}^{(2)}(T_{mn})(f)
$$

are given as follows,

$$
\begin{aligned}
\lambda_{p,q}\left(r_{F,F}^{(2)}(T_{mn})(f)\right) &= \sum_{s=0}^{m-1}\sum_{t=0}^{n-1} c_{s,t}\xi_p^s\eta_q^t \\
&= t_{0,0} + \sum_{j=1}^{m-1}(t_{j,0} + t_{j-m,0})\xi_p^j + \sum_{k=1}^{n-1}(t_{0,k} + t_{0,k-n})\eta_q^k \\
&\quad + \sum_{j=1}^{m-1}\sum_{k=1}^{n-1}(t_{j,k} + t_{j-m,k} + t_{j,k-n} + t_{j-m,k-n})\xi_p^j\eta_q^k \\
&= t_{0,0} + \sum_{j=-(m-1)}^{m-1} t_{j,0}\xi_p^j + \sum_{k=-(n-1)}^{n-1} t_{0,k}\eta_q^k \\
&\quad + \sum_{j=1}^{m-1}\sum_{k=1}^{n-1} t_{j,k}\xi_p^j\eta_q^k + \sum_{j=1}^{m-1}\sum_{k=1}^{n-1} t_{-j,k}\xi_p^{-j}\eta_q^k \\
&\quad + \sum_{j=1}^{m-1}\sum_{k=1}^{n-1} t_{j,-k}\xi_p^j\eta_q^{-k} + \sum_{j=1}^{m-1}\sum_{k=1}^{n-1} t_{-j,-k}\xi_p^{-j}\eta_q^{-k} \\
&= \sum_{j=-(m-1)}^{m-1}\sum_{k=-(n-1)}^{n-1} t_{j,k}\xi_p^j\eta_q^k \\
&= s_{(m-1)(n-1)}(f)(2\pi p/m, 2\pi q/n) \\
&= f * \widehat{D}_{(m-1)(n-1)}(2\pi p/m, 2\pi p/n)
\end{aligned}
$$

where $\widehat{D}_{(m-1)(n-1)}(x,y)$ is defined as in (3.5).

3.2.4 H_{mn}^{st}

We construct the BCCB preconditioner $H_{mn}^{st}(f)$ as follows

$$H_{mn}^{st}(f) = (F_m^* \otimes F_n^*)\Lambda_{mn}(F_m \otimes F_n).$$

Here Λ_{mn} is a diagonal matrix with the diagonal entries (the eigenvalues of $H_{mn}^{st}(f)$) given by

$$(\Lambda_{mn})_{p,q} = \frac{1}{st} \sum_{j=-s}^{s} \sum_{k=-t}^{t} (s - |j|)(t - |k|) t_{j,k} \xi_p^j \eta_q^k$$

$$= \frac{1}{st} \sum_{j=0}^{s-1} \sum_{k=0}^{t-1} s_{j,k}(f)(2\pi p/m, 2\pi q/n)$$

$$= f * \widehat{F_{st}}(2\pi p/m, 2\pi q/n)$$

where $\widehat{F_{st}}(x,y)$ is defined as in (3.7). We remark that $H_{mn}^{st}(f)$ is a block generalization of the preconditioner proposed by Huckle in [50], see also (1.14).

In fact, for any kernel $\widehat{C_{mn}}(x,y)$ defined on $[-\pi, \pi] \times [-\pi, \pi]$ we can design a BCCB preconditioner $C_{mn}(f)$ as follows. Let $\mathcal{M}_{F_m \otimes F_n}$ be the set of all BCCB matrices, see Section 2.1.4. Let $C_{mn}(f) \in \mathcal{M}_{F_m \otimes F_n}$ with the eigenvalues given by

$$\lambda_{p,q}(C_{mn}(f)) = f * \widehat{C_{mn}}(2\pi p/m, 2\pi p/n). \tag{3.9}$$

Then the first column of $C_{mn}(f)$ can be obtained by using (2.21). Once the first column of $C_{mn}(f)$ is found one can obtain the BCCB preconditioner $C_{mn}(f)$ easily.

3.3 Clustering properties

In this section, we discuss the convergence properties of preconditioned systems. We first introduce the following lemma, see [101].

Lemma 3.1 *Let* $f \in C_{2\pi \times 2\pi}$, *then*

$$|||f * \widehat{F_{mn}} - f|||_\infty \to 0 \tag{3.10}$$

when m, n *tend to infinity. Here* $\widehat{F_{mn}}$ *is defined by (3.7).*

By using Lemma 3.1 and Theorem 3.1, one can prove the following lemma easily.

Lemma 3.2 *Let* $f \in C_{2\pi \times 2\pi}$ *and* $\widehat{C_{mn}}$ *be a kernel such that*

$$|||f * \widehat{C_{mn}} - f|||_\infty \to 0$$

when m, n *tend to infinity. Let* $C_{mn}(f) \in \mathcal{M}_{F_m \otimes F_n}$ *with eigenvalues given by (3.9). Then for all* $\epsilon > 0$, *there exist* M *and* $N > 0$, *such that for all* $m > M$ *and* $n > N$, *at most* $O(m) + O(n)$ *eigenvalues of* $C_{mn}(f) - T_{mn}(f)$ *have absolute values larger than* ϵ.

Proof: We note that

$$T_{mn}(f) - C_{mn}(f)$$
$$= (T_{mn}(f) - c_{F,F}^{(2)}(T_{mn})) + (c_{F,F}^{(2)}(T_{mn}) - C_{mn}(f)).$$

Since

$$\|c_{F,F}^{(2)}(T_{mn}) - C_{mn}(f)\|_2 = \max_{p,q} |\lambda_{p,q}\left(c_{F,F}^{(2)}(T_{mn})\right) - \lambda_{p,q}(C_{mn}(f))|$$

$$= \max_{p,q} |(f * \widehat{F_{mn}} - f * \widehat{C_{mn}})(2\pi p/m, 2\pi p/n)|$$

$$\leq |||f * \widehat{F_{mn}} - f|||_\infty + |||f - f * \widehat{C_{mn}}|||_\infty,$$

the result follows by using Lemma 3.1, (3.10) and Theorem 3.1. $\square$

Lemma 3.3 *Let* $f \in C_{2\pi \times 2\pi}$ *with* $f_{\min} > 0$ *and* $\widehat{C_{mn}}$ *be a kernel such that*

$$f * \widehat{C_{mn}} \to f$$

uniformly on $[-\pi, \pi] \times [-\pi, \pi]$. *If* $C_{mn}(f) \in \mathcal{M}_{F_m \otimes F_n}$ *with eigenvalues given by (3.9), then for all* m, n *sufficiently large, we heve*

$$\lambda_{p,q}(C_{mn}(f)) \geq \frac{1}{2} f_{\min} > 0.$$

Proof: Similar to the proof of Lemma 3 in [29]. □

By using Lemmas 3.2 and 3.3, we have the following theorem.

Theorem 3.2 *Let* $f \in C_{2\pi \times 2\pi}$ *with* $f_{\min} > 0$ *and* $\widehat{C_{mn}}$ *be a kernel such that*

$$f * \widehat{C_{mn}} \to f$$

uniformly on $[-\pi, \pi] \times [-\pi, \pi]$. *Let* $C_{mn}(f) \in \mathcal{M}_{F_m \otimes F_n}$ *with eigenvalues given by (3.9). Then for all* $\epsilon > 0$ *there exist* M *and* $N > 0$ *such that for all* $m > M$ *and* $n > N$ *at most* $O(m) + O(n)$ *eigenvalues of* $I_{mn} - (C_{mn}(f))^{-1} T_{mn}(f)$ *have absolute values larger than* ϵ.

When the PCG method is applied to solving the system $T_{mn}(f)u = b$ with the preconditioner $C_{mn}(f)$, by Theorem 3.2, we can expect a fast convergence rate.

Chapter 4

Fast Algorithm for Tensor Structure

We study the solution of block systems $Tu = b$ by the PCG method, where

$$T = T_{(1)} \otimes T_{(2)} \otimes \cdots \otimes T_{(m)}$$

and $T_{(i)} \in \mathbb{R}^{n \times n}$, $i = 1, 2, \cdots, m$, are Toeplitz matrices. Preconditioners C and P, proposed in this chapter, are matrices that preserve the tensor structure of T. With a fast algorithm we show that both C and P are good preconditioners for solving block Toeplitz systems with tensor structure. Only $O(mn^m \log n)$ operations are required for the solution of preconditioned systems. An application in the inverse heat problem and numerical tests are also given.

4.1 Construction of preconditioner

Let us consider the block system

$$Tu = b$$

where

$$T = T_{(1)} \otimes T_{(2)} \otimes \cdots \otimes T_{(m)} \tag{4.1}$$

and $T_{(i)} \in \mathbb{R}^{n \times n}$, $i = 1, 2, \cdots, m$, are Toeplitz matrices. Each $T_{(i)}$ is generated by a function $f_i(x) \in \mathcal{C}_{2\pi}$, for $i = 1, 2, \cdots, m$. We first assume that $f_i(x) > 0$, for $i = 1, 2, \cdots, m$. Let

$$U \equiv \underbrace{F \otimes F \otimes \cdots \otimes F}_{m} \qquad (4.2)$$

where F is the $n \times n$ Fourier matrix. We then consider the preconditioner C which is defined to be the minimizer of $\|T - W\|_F$ over all $W \in \mathcal{M}_U$ where

$$\mathcal{M}_U = \{U^* \Lambda U \mid \Lambda \text{ is an } n^m \times n^m \text{ diagonal matrix}\}.$$

We then have the following theorem.

Theorem 4.1 *Let T be defined as in (4.1) and C be the minimizer of $\|T - W\|_F$ over all $W \in \mathcal{M}_U$ with U given by (4.2). Then*

(i) *C is uniquely determined by T and is given by*

$$C = C_{(1)} \otimes C_{(2)} \otimes \cdots \otimes C_{(m)}$$

where

$$C_{(i)} = c_F(T_{(i)}) = F^* \delta(F T_{(i)} F^*) F$$

is T. Chan's circulant preconditioner for $T_{(i)}$, $i = 1, 2, \cdots, m$, see Section 1.2.2.

(ii) *If T is Hermitian, then C is also Hermitian. Furthermore, we have*

$$\lambda_{\min}(T) \leq \lambda_{\min}(C) \leq \lambda_{\max}(C) \leq \lambda_{\max}(T).$$

In particular, if T is positive definite, then so is C.

The proof of Theorem 4.1 is straightforward by using Lemma 1.1, we therefore omit it.

If the generating functions $f_i(x)$ $(i = 1, 2, \cdots, m)$ have zeros, By Theorem 1.11, we know that the matrix T defined as in (4.1) is ill conditioned. We then consider the following preconditioner

$$P = P_{(1)} \otimes P_{(2)} \otimes \cdots \otimes P_{(m)}$$

where $P_{(i)}$ is the $\{\omega\}$-circulant preconditioner for $T_{(i)}$ proposed in Section 1.3.3.

4.2   Fast algorithm

For simplicity we start with $T = T_{(1)} \otimes T_{(2)}$. We first assume that $T_{(1)}$, $T_{(2)}$ are generated by two positive continuous functions $f_1(x)$, $f_2(x)$ respectively. For solving $Tu = b$, we use the PCG method with the preconditioner

$$C = C_{(1)} \otimes C_{(2)}$$

defined by Theorem 4.1 (i). The preconditioned system is given by

$$(C_{(1)}^{-1}T_{(1)} \otimes C_{(2)}^{-1}T_{(2)})u = \tilde{b} \tag{4.3}$$

where $\tilde{b} = C^{-1}b$. An algorithm for solving (4.3) is given as follows:

Algorithm 4.1

(i) Solve $(I_n \otimes C_{(2)}^{-1}T_{(2)})y = \tilde{b}$;

(ii) solve $(C_{(1)}^{-1}T_{(1)} \otimes I_n)u = y$.

In step (i) it is clear that the number of distinct eigenvalues of $I_n \otimes C_{(2)}^{-1}T_{(2)}$ is the same as the number of distinct eigenvalues of $C_{(2)}^{-1}T_{(2)}$. In view of Theorem 1.13 we note that for all $\epsilon > 0$, there exist $N, M > 0$ such that for all $n > N$, at most M distinct eigenvalues of the matrices $I_n \otimes C_{(2)}^{-1}T_{(2)} - I_{n^2}$ have absolute values larger than ϵ. Therefore the eigenvalues of the preconditioned matrix is clustered around 1. When the PCG method is applied to step (i), the convergence rate will be superlinear. The complexity of step (i) is equal to $O(n^2 \log n)$. For step (ii), with a similar analysis as we did for step (i), we need $O(n^2 \log n)$ operations to solve the preconditioned system in step (ii) by using the PCG method, and the convergence rate will be superlinear. Thus the total number of operations of our algorithm in solving the preconditioned system (4.3) is also of $O(n^2 \log n)$.

If $f_1(x)$, $f_2(x) \in C_{2\pi}$ have zeros we then consider to use the preconditioner of the form

$$P = P_{(1)} \otimes P_{(2)}$$

where $P_{(1)}$ and $P_{(2)}$ are $\{\omega\}$-circulant preconditioners proposed in Section 1.3.3. By using Algorithm 4.1 again and by noting the convergence results in Section 1.3.3, we know that only $O(n^2 \log n)$ operations are needed to

solve the preconditioned system by using the PCG method and the convergence rate will be superlinear. Therefore the total number of operations
of this algorithm is still of $O(n^2 \log n)$.

In general, for solving the system $Tu = b$ with

$$T = T_{(1)} \otimes T_{(2)} \otimes \cdots \otimes T_{(m)},$$

we first assume that each $T_{(i)}$ is generated by a function $f_i(x) > 0$ in $\mathcal{C}_{2\pi}$,
$i = 1, 2, \cdots, m$. By using the preconditioner

$$C = C_{(1)} \otimes C_{(2)} \otimes \cdots \otimes C_{(m)}$$

defined by Theorem 4.1 (i), the preconditioned system is given by

$$(C_{(1)}^{-1}T_{(1)} \otimes C_{(2)}^{-1}T_{(2)} \otimes \cdots \otimes C_{(m)}^{-1}T_{(m)})u = \tilde{b} \tag{4.4}$$

where $\tilde{b} = C^{-1}b$. An algorithm for solving (4.4) is given as follows:

Algorithm 4.2

(1) Solve $(I_n \otimes I_n \otimes \cdots \otimes I_n \otimes C_{(m)}^{-1}T_{(m)})y_{(m-1)} = \tilde{b}$;

(2) solve $(I_n \otimes I_n \otimes \cdots \otimes C_{(m-1)}^{-1}T_{(m-1)} \otimes I_n)y_{(m-2)} = y_{(m-1)}$;

....

(m) solve $(C_{(1)}^{-1}T_{(1)} \otimes I_n \otimes \cdots \otimes I_n \otimes I_n)u = y_{(1)}$.

With a similar analysis as we did for Algorithm 4.1, we note that the
total number of operations of our algorithm in solving the preconditioned
system (4.4) is of $O(mn^m \log n)$. If the generating functions $f_i(x)$ ($i =
1, 2, \cdots, m$) have zeros, one could consider using

$$P = P_{(1)} \otimes P_{(2)} \otimes \cdots \otimes P_{(m)}$$

instead of C.

4.3   Inverse heat problem

In this section, we consider the inverse heat problem in $\mathbb{R}^n$. For simplicity
we begin with the inverse heat problem in $\mathbb{R}$. The heat equation is given
by

$$\partial_t u(x,t) = \partial_x^2 u(x,t), \qquad \forall x \in \mathbb{R}, \quad t > 0,$$

where the initial values are given by $u(x,0) = f(x)$ for all $x \in \mathbb{R}$. It is
well known that if $f \in L^2(\mathbb{R})$, then

$$u(x,t) = \frac{1}{\sqrt{4\pi t}} \int_{-\infty}^{\infty} \exp\left(\frac{-(x-y)^2}{4t}\right) f(y)dy, \tag{4.5}$$

see [61].

The inverse heat problem in $\mathbb{R}$ is the problem of recovering the initial
data $f(y)$ for all $y \in \mathbb{R}$ when for some $t > 0$, $u(x,t)$ is given for all $x \in \mathbb{R}$.
We will restrict to the following class of functions, see [37] and [86].

Definition 4.1 *Let $h > 0$ be a constant. A function f is said to be in the
class $B(h)$, if it satisfies the following three conditions:*

(i) *f is an entire function;*

(ii) *on the real line $\mathbb{R}$, $f \in L^2(\mathbb{R})$;*

(iii) *in the complex plane $\mathbb{C}$,*

$$|f(z)| < K \exp\left(\frac{\pi|z|}{h}\right)$$

for some $K > 0$.

For $f \in B(h)$, we approximate $f(y)$ in (4.5) by

$$f(y) \approx \sum_{k=-\infty}^{\infty} f(kh)\operatorname{sinc}\left(\frac{y-kh}{h}\right)$$

where

$$\operatorname{sinc}(x) \equiv \begin{cases} \dfrac{\sin(\pi x)}{\pi x}, & x \neq 0, \\ \\ 1, & x = 0, \end{cases}$$

see [37] and [86]. Then (4.5) becomes

$$u(x,t) \approx \frac{1}{\sqrt{4\pi t}} \sum_{k=-\infty}^{\infty} f(kh) \int_{-\infty}^{\infty} \exp\left(-\frac{(x-y)^2}{4t}\right) \text{sinc}\left(\frac{y-kh}{h}\right) dy.$$

Let $x = x_j = jh$ and $t_0 = \frac{h}{2\pi}$. After some simplification, see [37], we finally have

$$u(x_j,t_0) \approx \frac{1}{2\pi} \sum_{k=-\infty}^{\infty} f(kh) \int_{-\pi}^{\pi} \exp\left(-\frac{\tau^2}{4\pi^2}\right) e^{i(k-j)\tau} d\tau$$

$$\equiv \sum_{k=-\infty}^{\infty} f(kh)\beta_{k-j}$$

where

$$\beta_p = \frac{1}{2\pi} \int_{-\pi}^{\pi} g(\tau) e^{ip\tau} d\tau,$$

for $p = 0, \pm1, \pm2, \cdots$, are the Fourier coefficients of the real-valued even function

$$g(\tau) \equiv \exp\left(-\frac{\tau^2}{4\pi^2}\right).$$

Clearly $g(\tau)$ is positive and in $C^\infty[-\pi,\pi]$. The function $g(\tau)$ can be extended continuously to a 2π-periodic function in $\mathbb{R}$. For a finite $n > 0$ we then have the discrete system:

$$u(x_j,t_0) = \sum_{k=-n}^{n} f(kh)\beta_{k-j},$$

or in matrix form

$$T\tilde{f} = \tilde{u}. \tag{4.6}$$

Here

$$T = \begin{pmatrix} \beta_0 & \beta_1 & \cdots & \beta_{2n} \\ \beta_1 & \beta_0 & \cdots & \beta_{2n-1} \\ \vdots & \vdots & \ddots & \vdots \\ \beta_{2n} & \beta_{2n-1} & \cdots & \beta_0 \end{pmatrix} \in \mathbb{R}^{(2n+1)\times(2n+1)}, \tag{4.7}$$

$$\tilde{f} = (f(-nh), f(-nh+h), \cdots, f(nh-h), f(nh))^T$$

and

$$\tilde{u} = (u(x_{-n},t_0), \cdots, u(x_0,t_0), \cdots, u(x_n,t_0))^T.$$

Given $\tilde{u}$ we can solve (4.6) to determine $\tilde{f}$ and use it to approximate $f(y)$ for all $y \in \mathbb{R}$. The inverse heat problem has now been converted into a problem of solving Toeplitz system (4.6).

In the n-dimensional case one can easily check that the discrete block Toeplitz system to be solved is of the following form:

$$\tilde{T}f = u \qquad (4.8)$$

where

$$\tilde{T} = \underbrace{T \otimes T \otimes \cdots \otimes T}_{n}$$

with T given by (4.7). We then precondition (4.8) with C or P proposed in Section 4.1. If the preconditioned system is solved by using Algorithm 4.2, we can obtain a superlinear convergence rate. Some numerical tests of this problem could be found in [15].

4.4   Numerical results

In this section we apply the PCG method with our fast algorithm proposed in Section 4.2 for solving block Toeplitz systems $Tu = b$, where $T = T_{(1)} \otimes T_{(2)}$. Each $T_{(i)}$ is generated by a function $f_i \in C_{2\pi}$ for $i = 1, 2$. The preconditioners used are C and P proposed in Section 4.1. For a comparison we also test the block algorithm proposed in Chapter 2.

We test the following six systems with different generating functions defined on $[-\pi, \pi]$. They are separated into two classes:

Positive continuous generating functions:

1. $f_1(x) = x^6 + 1, \quad f_2(y) = |y|^3 + 0.01;$
2. $f_1(x) = (\cos x)^2 + 0.1, \quad f_2(y) = |y|^5 + \pi;$
3. $f_1(x) = x^2 + \pi/2, \quad f_2(y) = y^4 + 1.$

Non-negative continuous generating functions with zeros:

4. $f_1(x) = (x^2 - 1)^2, \quad f_2(y) = y^2;$
5. $f_1(x) = |x|^3, \quad f_2(y) = |y|^3;$
6. $f_1(x) = x^4, \quad f_2(y) = y^4 + (\sin y)^2.$

Tables 4.1 to 4.6 show the number of iterations required for convergence with different choices of preconditioners. In these tables, iteration numbers more than 10,000 are denoted by '$\star$'. In all tests, we set

$w_n = -\pi + \pi/n$ for the construction of the preconditioner P, see Section 1.3.3. When the PCG method is applied to such kind of system, the stopping criteria is

$$\frac{\|r^{(k)}\|_2}{\|r^{(0)}\|_2} < 10^{-7}$$

where $r^{(k)}$ is the residual vector after k-th iterations. The right hand side vector b is chosen to be the vector of all ones and the zero vector is the initial guess.

In all tests, when we compare the fast algorithm with the block algorithm proposed in Chapter 2, the performance of the fast algorithm is better. In each step of the fast algorithm, the preconditioned systems with the preconditioner P converge at a rate that is independent of the size of matrices, while the preconditioned systems with the preconditioner C converge at a rate that increases slowly with the size of matrices. We note that the preconditioner C is not good enough for the last three systems which have generating functions with zeros on $[-\pi, \pi]$. This is the reason why we apply the preconditioner P to handle these ill conditioned systems. In all tables, 'Fast Algo.' means the fast algorithm and 'Block Algo.' means the block algorithm. As before, I means that no preconditioner is used. For solving both well conditioned and ill conditioned block Toeplitz systems by using the PCG method, the performances of the preconditioner P are more effective than those of the preconditioner C.

Table 4.1. Number of iterations for $f_1(x) = x^6 + 1$, $f_2(y) = |y|^3 + 0.01$

$m = n$	mn	I			C			P		
		Block Algo.	Fast Algo. (i)	(ii)	Block Algo.	Fast Algo. (i)	(ii)	Block Algo.	Fast Algo. (i)	(ii)
16	256	711	10	8	108	8	8	17	6	6
32	1024	5535	28	21	180	15	13	17	7	7
64	4096	$\star$	59	53	146	13	15	17	7	7
128	16384	$\star$	107	132	109	11	18	17	7	7
256	65536	$\star$	168	274	78	9	15	17	7	7

Table 4.2. Number of iterations for $f_1(x) = (\cos x)^2 + 0.1$, $f_2(y) = |y|^5 + \pi$

$m = n$	mn	I Block Algo.	Fast Algo. (i)	(ii)	C Block Algo.	Fast Algo. (i)	(ii)	P Block Algo.	Fast Algo. (i)	(ii)
16	256	57	4	8	16	4	8	9	2	5
32	1024	142	8	19	18	6	8	10	2	5
64	4096	191	16	33	17	5	7	10	2	5
128	16384	220	22	52	16	5	6	9	2	5
256	65536	223	21	64	14	4	6	9	2	5

Table 4.3. Number of iterations for $f_1(x) = x^2 + \pi/2$, $f_2(y) = y^4 + 1$

$m = n$	mn	I Block Algo.	Fast Algo. (i)	(ii)	C Block Algo.	Fast Algo. (i)	(ii)	P Block Algo.	Fast Algo. (i)	(ii)
16	256	108	8	8	19	6	8	11	4	5
32	1024	163	16	20	17	5	7	11	4	5
64	4096	190	19	36	16	5	7	11	4	5
128	16384	202	19	55	14	5	6	10	4	5
256	65536	203	19	67	14	4	6	10	4	5

Table 4.4. Number of iterations for $f_1(x) = (x^2 - 1)^2$, $f_2(y) = y^2$

$m = n$	mn	I Block Algo.	Fast Algo. (i)	(ii)	C Block Algo.	Fast Algo. (i)	(ii)	P Block Algo.	Fast Algo. (i)	(ii)
16	256	359	9	8	64	8	8	22	6	4
32	1024	2608	25	17	125	14	10	25	6	5
64	4096	$\star$	71	38	271	17	12	24	6	6
128	16384	$\star$	191	82	559	22	14	32	8	6
256	65536	$\star$	460	172	1260	27	17	28	8	6

Table 4.5. Number of iterations for $f_1(x) = |x|^3$, $f_2(y) = |y|^3$

$m = n$	mn	I			C			P		
		Block Algo.	Fast Algo. (i)	(ii)	Block Algo.	Fast Algo. (i)	(ii)	Block Algo.	Fast Algo. (i)	(ii)
16	256	271	8	8	56	8	8	14	6	6
32	1024	3433	22	22	206	13	13	19	7	7
64	4096	$\star$	60	60	1196	17	17	25	9	9
128	16384	$\star$	176	176	5860	24	24	33	9	9
256	65536	$\star$	562	562	$\star$	36	36	51	9	9

Table 4.6. Number of iterations for $f_1(x) = x^4$, $f_2(y) = y^4 + (\sin y)^2$

$m = n$	mn	I			C			P		
		Block Algo.	Fast Algo. (i)	(ii)	Block Algo.	Fast Algo. (i)	(ii)	Block Algo.	Fast Algo. (i)	(ii)
16	256	1166	10	9	146	8	8	22	6	6
32	1024	$\star$	33	24	843	16	14	31	8	6
64	4096	$\star$	113	64	4688	25	17	43	9	7
128	16384	$\star$	487	170	$\star$	44	22	51	10	7
256	65536	$\star$	2399	432	$\star$	136	29	70	12	7

Chapter 5

Block Toeplitz LS Problems

We study the constrained and weighted least squares (LS) problem

$$\min_x (b - Tx)^T W(b - Tx)/2$$

where

$$W = \text{diag}(\omega_1, \cdots, \omega_m)$$

with $\omega_1 \geq \cdots \geq \omega_m \geq 0$, and

$$T = \left(T_{(1)}, \cdots, T_{(k)}\right)^T$$

with Toeplitz blocks $T_{(l)} \in \mathbb{R}^{n \times n}$ for $l = 1, \cdots, k$. It is well known that this problem can be solved by solving the following block system

$$\begin{cases} M\lambda + Tx = b, \\ \\ T^T\lambda = 0, \end{cases}$$

where $M = W^{-1}$. We use the PCG method with circulant-like preconditioners for solving the system. The spectrum of the preconditioned matrix is clustered around one. We then have a fast convergence rate. For a literature on LS problems, we refer to [38, 41, 56, 42].

5.1 Introduction

The general constrained and weighted LS problem is defined as follows,

$$\min_x (b - Ax)^T W(b - Ax)/2 \tag{5.1}$$

where $A \in \mathbb{R}^{m \times n}$ with $m \geq n$, and

$$W = \mathrm{diag}(\omega_1, \cdots, \omega_m)$$

with

$$\omega_1 \geq \cdots \geq \omega_m \geq 0.$$

If W is invertible, we can define the following linear system

$$\begin{pmatrix} M & A \\ A^T & 0 \end{pmatrix} \begin{pmatrix} \lambda \\ x \end{pmatrix} = \begin{pmatrix} b \\ 0 \end{pmatrix} \tag{5.2}$$

where $M = W^{-1}$. Actually, the system (5.2) defines the whole class of constrained and weighted LS problems, see [41]. More precisely, we may assume that the first p diagonal entries in M are zero and the rest diagonal entries are non-zero. Then, the LS problem (5.1) could be reformulated as

$$\min_x (b_2 - A_{(2)}x)^T \tilde{W}(b_2 - A_{(2)}x)/2 \quad \text{such that} \quad A_{(1)}x = b_1. \tag{5.3}$$

Here $\tilde{W} \in \mathbb{R}^{(m-p) \times (m-p)}$ is a diagonal matrix with positive diagonal entries and

$$A = \begin{pmatrix} A_{(1)} \\ A_{(2)} \end{pmatrix}, \quad b = \begin{pmatrix} b_1 \\ b_2 \end{pmatrix}$$

with

$$A_{(1)} \in \mathbb{R}^{p \times n}, \quad A_{(2)} \in \mathbb{R}^{(m-p) \times n}, \quad b_1 \in \mathbb{R}^p, \quad b_2 \in \mathbb{R}^{(m-p)}.$$

We assume that the LS problem (5.3) has a unique solution.

We will study the solution of (5.2) where

$$M \equiv \mathrm{diag}(\mu_1, \cdots, \mu_m) \tag{5.4}$$

with $\mu_1 = \cdots = \mu_p = 0$ and $0 < \mu_{p+1} \leq \cdots \leq \mu_m \leq 1$, and

$$A = T \equiv \left(T_{(1)}, \cdots, T_{(k)} \right)^T \tag{5.5}$$

with symmetric Toeplitz blocks $T_{(l)} \in \mathbb{R}^{n \times n}$ for $l = 1, \cdots, k$. Here $m = kn$ with a positive integer k which is independent of n. Each $T_{(l)}$ is generated by a real-valued even function $f^{(l)}$ with diagonals given by

$$t_p^{(l)}(f^{(l)}) = \frac{1}{2\pi} \int_{-\pi}^{\pi} f^{(l)}(x) e^{-ipx} dx, \quad p = 0, \pm 1, \pm 2, \cdots.$$

Such kind of problem has been studied in [22] with $M = I_m$, the identity matrix. Some applications of the problem studied here could be found in [39]. In the following we assume that

$$0 < f_{\min}^{(l)} \le f^{(l)} \le f_{\max}^{(l)} < \infty \tag{5.6}$$

and that each $f^{(l)}$ is in the Wiener class, for $l = 1, \cdots, k$. We will use the PCG method with a circulant-like preconditioner for solving (5.2) where M is defined by (5.4) and $A = T$ is defined by (5.5). We show that the spectrum of the preconditioned matrix is clustered around one. When the PCG method is applied to solving the system, we can expect a fast convergence rate.

5.2   Construction of preconditioner

Let

$$S \equiv \begin{pmatrix} M & T \\ T^T & 0 \end{pmatrix} \tag{5.7}$$

with M defined by (5.4) and T defined by (5.5). We construct our preconditioner for S as follows. Let

$$E \equiv \begin{pmatrix} c_F(M) & C \\ C^T & 0 \end{pmatrix} \tag{5.8}$$

where $c_F(M)$ is defined as in (1.10) and

$$C = \left(c_F(T_{(1)}), \cdots, c_F(T_{(k)}) \right)^T.$$

In fact, $c_F(M)$ is given by

$$c_F(M) = \mathrm{diag}(\mu^{(m)}, \cdots, \mu^{(m)}) = \mu^{(m)} I$$

with $\mu^{(m)} = \left(\sum_{i=p+1}^{m} \mu_i \right) / m$. Since $\{\mu_i\}$ is a monotonic increasing sequence bounded by 1, we have

$$\lim_{m \to \infty} \mu^{(m)} = \lim_{n \to \infty} \mu^{(kn)} = \lim_{i \to \infty} \mu_i = a > 0. \tag{5.9}$$

We now consider the spectral properties of E defined by (5.8). Since every circulant matrices can be diagonalized by the Fourier matrix F, we have the following decomposition by Lemma 1.1,

$$E = \begin{pmatrix} c_F(M) & C \\ C^T & 0 \end{pmatrix} = (I_{k+1} \otimes F^*) \begin{pmatrix} \mu^{(m)}I & \Delta \\ \Delta^T & 0 \end{pmatrix} (I_{k+1} \otimes F) \tag{5.10}$$

where

$$\Delta \equiv \left(\text{diag}(FT_{(1)}F^*), \cdots, \text{diag}(FT_{(k)}F^*) \right)^T.$$

From (5.10), we know that the spectrum of E is equal to the spectrum of

$$B \equiv \begin{pmatrix} \mu^{(m)}I & \Delta \\ \Delta^T & 0 \end{pmatrix}. \tag{5.11}$$

We will find the eigenvalues of B. Note that the matrix B is a $(k+1) \times (k+1)$ block matrix with $n \times n$ diagonal blocks. One can easily obtain the following matrix after a permutation,

$$\tilde{B} \equiv P^* BP = \begin{pmatrix} B_{1,1} & 0 & \cdots & 0 \\ 0 & B_{2,2} & \ddots & \vdots \\ \vdots & \ddots & \ddots & 0 \\ 0 & \cdots & 0 & B_{n,n} \end{pmatrix} \tag{5.12}$$

where P is the permutation matrix defined as in (2.6). In fact, the matrix $\tilde{B}$ is an $n \times n$ block diagonal matrix with blocks $B_{j,j} \in \mathbb{R}^{(k+1) \times (k+1)}$ for $j = 1, \cdots, n$. Each block $B_{j,j}$ is given by

$$B_{j,j} = \begin{pmatrix} \mu^{(m)} & 0 & \cdots & 0 & \lambda_1^{(j)} \\ 0 & \mu^{(m)} & \ddots & \vdots & \lambda_2^{(j)} \\ \vdots & \ddots & \ddots & 0 & \vdots \\ 0 & \cdots & 0 & \mu^{(m)} & \lambda_k^{(j)} \\ \lambda_1^{(j)} & \lambda_2^{(j)} & \cdots & \lambda_k^{(j)} & 0 \end{pmatrix}$$

where $\lambda_l^{(j)}$ is the j-th diagonal entry from the l-th diagonal block $\mathrm{diag}(FT_{(l)}F^*$ of Δ. We obtain the eigenvalues of $B_{j,j}$ by solving the equation

$$\det(\lambda I_{k+1} - B_{j,j}) \equiv (\lambda - \mu^{(m)})^{k-1}(\lambda^2 - \mu^{(m)}\lambda - \sum_{l=1}^{k}(\lambda_l^{(j)})^2) = 0.$$

Hence, the eigenvalues of $B_{j,j}$ are given by

$$\begin{cases} \lambda_1(B_{j,j}) = \cdots = \lambda_{k-1}(B_{j,j}) = \mu^{(m)}, \\[2mm] \lambda_k(B_{j,j}) = \dfrac{1}{2}\left(\mu^{(m)} + \sqrt{(\mu^{(m)})^2 + 4\sum_{l=1}^{k}(\lambda_l^{(j)})^2}\right), \\[3mm] \lambda_{k+1}(B_{j,j}) = \dfrac{1}{2}\left(\mu^{(m)} - \sqrt{(\mu^{(m)})^2 + 4\sum_{l=1}^{k}(\lambda_l^{(j)})^2}\right). \end{cases}$$

Thus the spectrum of $\tilde{B}$ (and hence E) is given as follows,

$$\begin{cases} \mu^{(m)} \quad \text{with } (k-1)n \text{ multiplicity}, \\[2mm] \dfrac{1}{2}\left(\mu^{(m)} + \sqrt{(\mu^{(m)})^2 + 4\sum_{l=1}^{k}(\lambda_l^{(j)})^2}\right), \quad j = 1, \cdots, n, \\[3mm] \dfrac{1}{2}\left(\mu^{(m)} - \sqrt{(\mu^{(m)})^2 + 4\sum_{l=1}^{k}(\lambda_l^{(j)})^2}\right), \quad j = 1, \cdots, n. \end{cases}$$

Lemma 5.1 *Under the conditions (5.6) and (5.9), $\|E\|_2$ and $\|E^{-1}\|_2$ are uniformly bounded as n increases.*

Proof: When $n \to \infty$, by (5.6), (5.9), Theorem 1.10 and Lemma 1.1 (ii), one can find that all the eigenvalues are uniformly bounded:

$$\lim_{m=kn\to\infty} \mu^{(m)} = a > 0;$$

$$0 < \frac{1}{2}\left(a + \sqrt{a^2 + 4\sum_{l=1}^{k}(f_{\min}^{(l)})^2}\right)$$

$$\leq \lim_{m=kn\to\infty} \frac{1}{2}\left(\mu^{(m)} + \sqrt{(\mu^{(m)})^2 + 4\sum_{l=1}^{k}(\lambda_l^{(j)})^2}\right)$$

$$\leq \frac{1}{2}\left(a + \sqrt{a^2 + 4\sum_{l=1}^{k}(f_{\max}^{(l)})^2}\right) < \infty, \quad j = 1, \cdots, n,$$

and

$$-\infty < \frac{1}{2}\left(a - \sqrt{a^2 + 4\sum_{l=1}^{k}(f_{\max}^{(l)})^2}\right)$$

$$\leq \lim_{m=kn\to\infty} \frac{1}{2}\left(\mu^{(m)} - \sqrt{(\mu^{(m)})^2 + 4\sum_{l=1}^{k}(\lambda_l^{(j)})^2}\right)$$

$$\leq \frac{1}{2}\left(a - \sqrt{a^2 + 4\sum_{l=1}^{k}(f_{\min}^{(l)})^2}\right) < 0, \quad j = 1, \cdots, n. \quad \square$$

5.3 Spectrum of preconditioned matrix

We consider the preconditioned matrix $E^{-1}S$ with S defined by (5.7) and E defined by (5.8). Since

$$E^{-1}S = I + E^{-1}(S - E) \tag{5.13}$$

and $\|E^{-1}\|_2$ is uniformly bounded, we want to show that the eigenvalues of

$$S - E = \begin{pmatrix} M - c_F(M) & T - C \\ T^T - C^T & 0 \end{pmatrix}$$

are clustered around zero. For the matrix

$$c_F(M) - M = \text{diag}(\mu^{(m)}, \cdots, \mu^{(m)}, \mu^{(m)} - \mu_{p+1}, \cdots, \mu^{(m)} - \mu_m),$$

we have from (5.9) that for all $\epsilon > 0$ there exist N_1 and M_1 such that for all $n > N_1$ and $j > M_1$,

$$|\mu^{(m)} - \mu_j| = |\mu^{(kn)} - \mu_j| < \epsilon.$$

We therefore have

$$c_F(M) - M = U_{(1)} + V_{(1)} \tag{5.14}$$

with

$$U_{(1)} = \operatorname{diag}(\mu^{(m)}, \cdots, \mu^{(m)}, \mu^{(m)} - \mu_{p+1}, \cdots, \mu^{(m)} - \mu_{M_1}, 0, \cdots, 0)$$

and

$$V_{(1)} = \operatorname{diag}(0, \cdots, 0, \mu^{(m)} - \mu_{M_1+1}, \cdots, \mu^{(m)} - \mu_m).$$

Note that

$$\operatorname{rank}(U_{(1)}) \leq M_1 \quad \text{and} \quad \|V_{(1)}\|_1 \leq \epsilon.$$

For the matrix $T - C$, we have the following lemma. The proof of the lemma is similar to that of Theorem 2.5.

Lemma 5.2 *Under the condition (5.6), then for all $\epsilon > 0$, there exist N_2 and M_2 such that for all $n > N_2$,*

$$T - C = U_{(2)} + V_{(2)} \tag{5.15}$$

where $\operatorname{rank}(U_{(2)}) \leq M_2$ and $\|V_{(2)}\|_1 \leq \epsilon$.

By using (5.14) and (5.15) we have

$$S - E = U_{(3)} + V_{(3)} \tag{5.16}$$

with

$$U_{(3)} = \begin{pmatrix} U_{(1)} & U_{(2)} \\ U_{(2)}^T & 0 \end{pmatrix} \quad \text{and} \quad V_{(3)} = \begin{pmatrix} V_{(1)} & V_{(2)} \\ V_{(2)}^T & 0 \end{pmatrix}.$$

Therefore,

$$\operatorname{rank}(U_{(3)}) \leq M_3 = M_1 + M_2 \quad \text{and} \quad \|V_{(3)}\|_1 \leq 2\epsilon$$

when $n > N_3 = \max(N_1, N_2)$. Since $V_{(3)}$ is symmetric we have $\|V_{(3)}\|_1 = \|V_{(3)}\|_\infty$. Thus

$$\|V_{(3)}\|_2 \leq (\|V_{(3)}\|_1 \cdot \|V_{(3)}\|_\infty)^{1/2} \leq 2\epsilon.$$

From (5.13) and (5.16) we have

$$E^{-1}S = I + U_{(4)} + V_{(4)} \tag{5.17}$$

where

$$U_{(4)} = E^{-1}U_{(3)} \quad \text{and} \quad V_{(4)} = E^{-1}V_{(3)}$$

with

$$\text{rank}(U_{(4)}) \leq M_3 \quad \text{and} \quad \|V_{(4)}\|_2 \leq \epsilon_0 = 2\epsilon\|E^{-1}\|_2$$

when n is sufficiently large.

Let

$$y = \begin{pmatrix} \lambda \\ x \end{pmatrix} \quad \text{and} \quad d = \begin{pmatrix} b \\ 0 \end{pmatrix}.$$

It is clear that the solution of $Sy = d$ with S defined as in (5.7) is the same as the solution of the normalized preconditioned system

$$(E^{-1}S)^T(E^{-1}S)y = (E^{-1}S)^T E^{-1}d. \tag{5.18}$$

From (5.17) we have

$$(E^{-1}S)^T(E^{-1}S) - I = U + V$$

where

$$U \equiv U_{(4)}^T(I + U_{(4)} + V_{(4)}) + (I + V_{(4)}^T)U_{(4)}$$

and

$$V \equiv V_{(4)} + V_{(4)}^T + V_{(4)}^T V_{(4)}$$

with

$$\text{rank}(U) \leq 2M_3 \quad \text{and} \quad \|V\|_2 \leq 2\epsilon_0 + \epsilon_0^2$$

when n is sufficiently large. Thus we have the following theorem.

Theorem 5.1 *Under the conditions (5.6) and (5.9), the spectrum of the normalized preconditioned matrix $(E^{-1}S)^T(E^{-1}S)$ is clustered around one when n is sufficiently large.*

Hence if the PCG method is applied to solving the normalized preconditioned system (5.18), we can expect a fast convergence rate.

5.4   Convergence rate and operation cost

In this section, we analyse the convergence rate and the operation cost of our algorithm. We first show that the method converges in at most $O(\alpha \log n + 1)$ steps if $\sigma_{\min}(S) = O(n^{-\alpha})$ where $\sigma_{\min}(S)$ is the smallest singular value of S and $\alpha > 0$. By Theorem 1.9, if the eigenvalues λ_j of

$$G \equiv (E^{-1}S)^T (E^{-1}S)$$

are ordered such that

$$0 < \lambda_1 \leq \ldots \leq \lambda_p \leq b_1 \leq \lambda_{p+1} \leq \ldots \leq \lambda_{v-q} \leq b_2 \leq \lambda_{v-q+1} \leq \ldots \leq \lambda_v,$$

then the error vector $y - y^{(k)}$ after the k-th iteration of the PCG method when applied to the system (5.18) is given by

$$\frac{\|y - y^{(k)}\|}{\|y - y^{(0)}\|} \leq 2 \left(\frac{b-1}{b+1} \right)^{k-p-q} \cdot \max_{\lambda \in [b_1, b_2]} \prod_{j=1}^{p} \left(\frac{\lambda - \lambda_j}{\lambda_j} \right)$$

where $\|\cdot\|$ is the energy norm and $b \equiv (b_2/b_1)^{1/2} \geq 1$. By Theorem 5.1, we can choose $b_1 = 1 - \epsilon$ and $b_2 = 1 + \epsilon$ ($\epsilon > 0$). Then p and q are constants that depend only on ϵ but not on the size of the system. By choosing $\epsilon < 1$, we have by Corollary 1.1,

$$\frac{\|y - y^{(k)}\|}{\|y - y^{(0)}\|} \leq 2 \left(\frac{1+\epsilon}{\lambda_1} \right)^p \epsilon^{k-p-q}. \tag{5.19}$$

In order to estimate (5.19), we note that by Lemma 5.1,

$$\|G^{-1}\|_2 \leq \|E\|_2^2 \|S^{-1}\|_2^2 = \|E\|_2^2 \sigma_{\min}^{-2}(S) \leq c_1 n^{2\alpha}$$

for some $c_1 > 0$ independent of n, and

$$\lambda_1 = \|G^{-1}\|_2^{-1} \geq c_2 n^{-2\alpha} \tag{5.20}$$

where $c_2 = c_1^{-1}$. Substituting (5.20) into (5.19), we have

$$\frac{\|y - y^{(k)}\|}{\|y - y^{(0)}\|} \leq c^p n^{2p\alpha} \epsilon^{k-p-q}$$

where $c > 0$ is independent of n. Given arbitrary tolerance $\tau > 0$, an upper bound for the number of iterations required to make

$$\frac{\|y - y^{(k)}\|}{\|y - y^{(0)}\|} \leq \tau$$

is given by

$$k_0 = p + q - \frac{p \log c + 2\alpha p \log n - \log \tau}{\log \epsilon} = O(\alpha \log n + 1). \qquad (5.21)$$

Finally, we would like to consider the operation cost of solving the normalized preconditioned system (5.18) by the PCG method. We first recall that in each iteration of the method, we have to compute the matrix-vector multiplication Sv for some vector v and solve the system $Eu = w$ for some vector w. The matrix-vector multiplication

$$Sv = \begin{pmatrix} M & T \\ T^T & 0 \end{pmatrix} v = \begin{pmatrix} M & T \\ T^T & 0 \end{pmatrix} \begin{pmatrix} v_1 \\ v_2 \end{pmatrix} = \begin{pmatrix} Mv_1 + Tv_2 \\ T^T v_1 \end{pmatrix}$$

requires $O(kn \log n)$ operations for Tv_2 and $T^T v_1$ by using the algorithm proposed in Section 1.1.4, and $O(kn)$ operations for Mv_1. For solving the system $Eu = w$, we have by (5.10)–(5.12),

$$u = E^{-1}w = (I \otimes F^*) \begin{pmatrix} \mu^{(m)} I & \Delta \\ \Delta^T & 0 \end{pmatrix}^{-1} (I \otimes F)w$$

$$= (I \otimes F^*)B^{-1}(I \otimes F)w = (I \otimes F^*)(P\tilde{B}^{-1}P^*)(I \otimes F)w.$$

Since

$$\tilde{B}^{-1} = \begin{pmatrix} B_{1,1}^{-1} & 0 & \cdots & 0 \\ 0 & B_{2,2}^{-1} & \ddots & \vdots \\ \vdots & \ddots & \ddots & 0 \\ 0 & \cdots & 0 & B_{n,n}^{-1} \end{pmatrix},$$

only $O(k+1)$ operations are required for each $B_{j,j}^{-1}$, $j = 1, \cdots, n$, by using Gaussian elimination and thus $O(n(k+1))$ operations for $\tilde{B}^{-1}$. We need $O((k+1)n \log n)$ operations for $(I \otimes F)w$ by using FFTs. Thus, the cost per iteration is $O((k+1)n \log n)$. By (5.21), we conclude that the total complexity of our algorithm is $O(\alpha n \log^2 n + n \log n)$ if $\sigma_{\min}(S) = O(n^{-\alpha})$ with $\alpha > 0$.

Chapter 6

Block $\{\omega\}$-Circulant Preconditioners

We introduce two important theorems in spectral analysis. From spectral analysis, we note that the convergence rate of the conjugate gradient method for solving ill conditioned Toeplitz systems is slow usually. To deal with such kind of system, an $\{\omega\}$-circulant preconditioner is proposed in [71], see also Section 1.3.3. It was showed that the convergence rate of the PCG method is superlinear and the total operation cost is $O(n \log n)$, where n is the size of systems. In this chapter we generalize this preconditioner for BTTB systems. We show that block $\{\omega\}$-circulant preconditioners can work efficiently for ill conditioned BTTB systems. A numerical comparison between block $\{\omega\}$-circulant preconditioners and BCCB preconditioners is also given.

6.1   Spectral analysis

As before, BTTB systems $T_{mn}(f)u = b$ are assumed to be generated by a function $f(x, y) \in C_{2\pi \times 2\pi}$ with diagonals given by

$$t_{j,k}(f) \equiv t_k^{(j)}(f)$$
$$= \frac{1}{4\pi^2} \int_{-\pi}^{\pi} \int_{-\pi}^{\pi} f(x, y) e^{-i(jx+ky)} dx dy, \ j, k = 0, \pm 1, \pm 2, \cdots.$$

We begin by introducing two important theorems of spectral properties proposed by Serra in [77]. The theorems give the relations between the values of $f(x, y)$ and the eigenvalues of $T_{mn}(f)$. For more remarkable works on spectral analysis, we refer to [10, 79, 93, 94, 98]. The following theorem [77] is an improvement on Theorem 2.4.

Theorem 6.1 *Let* $f \in C_{2\pi \times 2\pi}$ *with the minimum* $f_{\min}$ *and maximum* $f_{\max}$. *If* $f_{\min} < f_{\max}$, *then for all* $m, n > 0$,

$$f_{\min} < \lambda_i(T_{mn}(f)) < f_{\max}, \quad i = 1, \cdots, mn,$$

where $\lambda_i(T_{mn})$ *is the i-th eigenvalue of* T_{mn}. *In particular, if* $f \geq 0$, *then* $T_{mn}(f)$ *are positive definite for all* m, n.

Proof:

We only prove that $f_{\min} < \lambda_i(T_{mn}(f))$. By Theorem 2.4 we know that $f_{\min} \leq \lambda_{\min}(T_{mn})$. By contradiction, let $\lambda_{\min}(T_{mn}) = f_{\min}$ and

$$u = (u_0^{(0)}, u_1^{(0)}, \cdots, u_{n-1}^{(0)}, u_0^{(1)}, \cdots, u_0^{(m-1)}, \cdots, u_{n-1}^{(m-1)})^T \in \mathbb{C}^{mn}$$

be the corresponding eigenvector with $u^* u = 1$. Then

$$T_{mn}(f)u = f_{\min} u.$$

By multiplying by u^* we obtain

$$u^* T_{mn}(f)u - f_{\min} = 0,$$

i.e.,

$$\frac{1}{4\pi^2} \int_{-\pi}^{\pi} \int_{-\pi}^{\pi} |p(z, w)|^2 (f(x, y) - f_{\min}) dx dy = 0,$$

where

$$p(z, w) \equiv \sum_{j=0}^{m-1} \sum_{k=0}^{n-1} u_k^{(j)} z^j w^k \qquad (6.1)$$

with $z \equiv e^{-ix}$, $w \equiv e^{-iy}$ and $(x, y) \in [-\pi, \pi] \times [-\pi, \pi]$. Let $\epsilon > 0$ such that the set

$$A(\epsilon) = \{(x, y) \in [-\pi, \pi] \times [-\pi, \pi] \mid f(x, y) - f_{\min} \geq \epsilon\}$$

has a positive Lebesgue measure. This is possible because of the continuity of f. Now, we have

$$
\begin{aligned}
0 &= \frac{1}{4\pi^2} \int_{-\pi}^{\pi} \int_{-\pi}^{\pi} |p(z,w)|^2 (f(x,y) - f_{\min}) dx dy \\
&\geq \epsilon \int_{A(\epsilon)} |p(z,w)|^2 dx dy \geq 0
\end{aligned}
$$

which implies

$$
\int_{A(\epsilon)} |p(z,w)|^2 dx dy = 0.
$$

Since the Lebesgue measure of $A(\epsilon)$ is positive, we have $p(z,w) \equiv 0$. Consequently we have $u = 0$, which is impossible. Thus,

$$
f_{\min} < \lambda_{\min}(T_{mn}). \qquad \square
$$

Furthermore, we have the following theorem [77],

Theorem 6.2 *Let $f \in C_{2\pi \times 2\pi}$ with $f_{\min} < f_{\max}$. We have*

$$
\lim_{m,n \to \infty} \lambda_{\max}(T_{mn}) = f_{\max} \tag{6.2}
$$

and

$$
\lim_{m,n \to \infty} \lambda_{\min}(T_{mn}) = f_{\min}. \tag{6.3}
$$

Proof:
   We only prove (6.2). For (6.3) it is possible to use the same arguments. First of all we show that

$$
\sup_{m,n} \lambda_{\max}(T_{mn}(f)) = f_{\max}.
$$

In view of Theorems 2.4 and 6.1, the eigenvalues of $T_{mn}(f)$ are bounded by $f_{\max}$ and therefore

$$
\sup_{m,n} \lambda_{\max}(T_{mn}(f)) \leq f_{\max}.
$$

Now, by contradiction, let

$$\sup_{m,n} \lambda_{\max}(T_{mn}(f)) = A \in (f_{\min}, f_{\max}),$$

then for any real number $\theta \in (A, f_{\max})$ we note that $\theta I - T_{mn}(f)$ is a positive definite matrix for any m, n. Hence we have

$$u^*(\theta I - T_{mn}(f))u = \frac{1}{4\pi^2} \int_{-\pi}^{\pi} \int_{-\pi}^{\pi} |p(z,w)|^2 (\theta - f(x,y)) dx dy > 0 \quad (6.4)$$

where

$$u = (u_0^{(0)}, u_1^{(0)}, \cdots, u_{n-1}^{(0)}, u_0^{(1)}, \cdots, u_0^{(m-1)}, \cdots, u_{n-1}^{(m-1)})^T$$

with $u^*u = 1$, and $p(z,w)$ is defined as in (6.1). Let

$$A_0^+ = \{(x,y) \in [-\pi, \pi] \times [-\pi, \pi] \mid \theta - f(x,y) \geq 0\}$$

and

$$A^- = \{(x,y) \in [-\pi, \pi] \times [-\pi, \pi] \mid \theta - f(x,y) < 0\}.$$

We notice that the Lebesque measure of A^- is greater than zero because of the continuity of f and the relation $\theta < f_{\max}$. Thus, (6.4) becomes

$$u^*(\theta I - T_{mn}(f))u = \frac{1}{4\pi^2} \int_{A_0^+} |p(z,w)|^2 (\theta - f(x,y)) dx dy$$

$$+ \frac{1}{4\pi^2} \int_{A^-} |p(z,w)|^2 (\theta - f(x,y)) dx dy$$

for any bivariate polynomial p defined on $\mathbb{T} \times \mathbb{T}$ where $\mathbb{T}$ is the unit circle in $\mathbb{C}$.

Now we consider a non-negative continuous function $g(x,y) \geq 0$ such that $g(x,y) = h(x)k(y)$ for any $(x,y) \in [-\pi, \pi] \times [-\pi, \pi]$. Here $h(x), k(y)$ are continuous functions on $[-\pi, \pi]$ with

$$h(\hat{x}) = 1, \quad k(\hat{y}) = 1$$

at an interior point $(\hat{x}, \hat{y})$ of A^- and

$$h(x) = 0 \quad \text{when } |x - \hat{x}| > \delta,$$

$$k(y) = 0 \quad \text{when } |y - \hat{y}| > \delta.$$

We can choose δ so that

$$[(\hat{x} - \delta, \hat{x} + \delta) \times (\hat{y} - \delta, \hat{y} + \delta)] \subseteq \left[A^{-} \cap [(-\pi, \pi) \times (-\pi, \pi)]\right].$$

Now $h(x), k(y)$ are two continuous real-valued non-negative functions of period 2π. By Weierstrass Approximation Theorem [72], for any $\epsilon > 0$, there exist two non-negative trigonometric polynomials $s(x), t(y)$ such that

$$|||h - s|||_{\infty} < \epsilon \quad \text{and} \quad |||k - t|||_{\infty} < \epsilon.$$

Moreover, by the Representation Theorem in [91] we note that any non-negative trigonometric polynomial can be expressed as the square of absolute value of a complex ordinary polynomial, i.e., $s = |p_1|^2$ and $t = |p_2|^2$ where $p_1 \equiv p_1(e^{-ix})$ and $p_2 \equiv p_2(e^{-iy})$. Thus we have

$$|||h - |p_1|^2|||_{\infty} < \epsilon \quad \text{and} \quad |||k - |p_2|^2|||_{\infty} < \epsilon.$$

Therefore, if $\tilde{u} \in \mathbb{C}^{\tilde{m}\tilde{n}}$ is the vector related to the coefficients of $p \equiv p_1(e^{-ix})p_2(e^{-iy})$, by setting

$$\theta(x, y) \equiv \theta - f(x, y),$$

we have

$$0 < 4\pi^2 (\tilde{u})^* (\theta I - T_{\tilde{m}\tilde{n}}(f))\tilde{u}$$

$$= \int_{-\pi}^{\pi} \int_{-\pi}^{\pi} (|p_1(z)|^2 - h(x))|p_2(w)|^2 \theta(x, y) dx dy$$

$$+ \int_{-\pi}^{\pi} \int_{-\pi}^{\pi} h(x)|p_2(w)|^2 \theta(x, y) dx dy$$

$$\leq \int_{-\pi}^{\pi} \int_{-\pi}^{\pi} (|p_1(z)|^2 - h(x))(k(y) + \epsilon)\theta(x, y) dx dy$$

$$+ \int_{-\pi}^{\pi} \int_{-\pi}^{\pi} h(x)(k(y) + \epsilon)\theta(x, y) dx dy$$

$$\leq \epsilon a + \int_{A^{-}} h(x)k(y)\theta(x, y) dx dy \equiv k(\epsilon),$$

where a is a positive constant. It follows that

$$\lim_{\epsilon \to 0} k(\epsilon) = \int_{A^-} g(x,y)(\theta - f(x,y))dxdy < 0$$

which is clearly impossible. We therefore have

$$\sup_{m,n} \lambda_{\max}(T_{mn}(f)) = f_{\max}.$$

Moreover for any $m \geq \tilde{m}$ and $n \geq \tilde{n}$, we observe that $T_{\tilde{m}\tilde{n}}(f)$ is a principal submatrix of $T_{mn}(f)$ and therefore by using Weyl's Theorem and Theorem 6.1, we have

$$f_{\max} > \lambda_{\max}(T_{mn}(f)) \geq \lambda_{\max}(T_{\tilde{m}\tilde{n}}(f)) > \theta,$$

i.e., $\lambda_{\max}(T_{mn}(f))$ is a bounded non-decreasing function of m and n. Consequently, we have

$$\lim_{m,n \to \infty} \lambda_{\max}(T_{mn}(f)) = \sup_{m,n} \lambda_{\max}(T_{mn}(f)) = f_{\max}. \qquad \square$$

From Theorems 6.1 and 6.2 we first note that if $f(x,y)$ is non-negative, then $T_{mn}(f)$ is always positive definite. We also know that when $f(x,y)$ vanishes at some points $(x_0, y_0) \in [-\pi, \pi] \times [-\pi, \pi]$, then the condition number $\kappa(T_{mn})$ of T_{mn} is unbounded as m, n tend to infinity, i.e., T_{mn} is ill conditioned. Since BCCB preconditioners proposed in Chapter 2 do not work well for ill conditioned systems, see numerical tests in Section 6.4, we then construct a block $\{\omega\}$-circulant preconditioner.

6.2 Construction of preconditioner

Let us consider the BTTB system

$$T_{mn}(f)x = b \tag{6.5}$$

where $T_{mn}(f)$ is generated by a function $f \geq 0$ in $C_{2\pi \times 2\pi}$ having a finite number of zeros. By Theorem 6.1, $T_{mn}(f)$ is positive definite. Let $u \in \mathbb{C}^{mn}$ be given by

$$u = (u_0^{(0)}, u_1^{(0)}, \cdots, u_{n-1}^{(0)}, u_0^{(1)}, \cdots, u_0^{(m-1)}, \cdots, u_{n-1}^{(m-1)})^T$$

with $u^*u = 1$. Then we have

$$u^*T_{mn}u = \frac{1}{4\pi^2} \int_{-\pi}^{\pi} \int_{-\pi}^{\pi} \left| \sum_{j=0}^{m-1} \sum_{k=0}^{n-1} u_k^{(j)} e^{-i(jx+ky)} \right|^2 f(x,y) dx dy. \qquad (6.6)$$

Now, we are going to construct a positive definite preconditioner $A_{mn}(f)$ for $T_{mn}(f)$ such that the preconditioned matrix $A_{mn}(f)^{-1}T_{mn}(f)$ has clustered spectrum.

Let grid points $(x_r, y_s) \in [-\pi, \pi] \times [-\pi, \pi]$ be given by

$$\begin{cases} x_r = \dfrac{2\pi r}{m} + \xi, \\[2mm] y_s = \dfrac{2\pi s}{n} + \eta, \end{cases}$$

for $r = 0, 1, \cdots, m-1$ and $s = 0, 1, \cdots, n-1$, where

$$(\xi, \eta) \in [-\pi, -\pi + \frac{2\pi}{m}) \times [-\pi, -\pi + \frac{2\pi}{n}).$$

Since f has only a finite number of zeros, we can choose a suitable pair $(\xi, \eta) \in [-\pi, -\pi + \frac{2\pi}{m}) \times [-\pi, -\pi + \frac{2\pi}{n})$ such that

$$f(x_r, y_s) > 0. \qquad (6.7)$$

Note that the choice of grid points requires some preparatory information about the zeros of $f(x, y)$.

To approximate the integral on the right-hand side of (6.6) by using the trapezoidal rule with respect to above grid points, we obtain

$$u^* T_{mn}(f) u$$

$$= \frac{1}{4\pi^2} \int_{-\pi}^{\pi} \int_{-\pi}^{\pi} \left| \sum_{j=0}^{m-1} \sum_{k=0}^{n-1} u_k^{(j)} e^{-i(jx+ky)} \right|^2 f(x,y) \, dx \, dy$$

$$\approx \frac{1}{mn} \sum_{r=0}^{m-1} \sum_{s=0}^{n-1} \left| \sum_{j=0}^{m-1} \sum_{k=0}^{n-1} u_k^{(j)} e^{-i(jx_r+ky_s)} \right|^2 f(x_r, y_s)$$

$$= \sum_{r=0}^{m-1} \sum_{s=0}^{n-1} f(x_r, y_s)$$

$$\times \frac{1}{\sqrt{mn}} \left(\sum_{j=0}^{m-1} \sum_{k=0}^{n-1} \bar{u}_k^{(j)} e^{-2\pi i(jr/m+ks/n)} e^{-i(j\xi+k\eta)} \right)$$

$$\times \frac{1}{\sqrt{mn}} \left(\sum_{j=0}^{m-1} \sum_{k=0}^{n-1} u_k^{(j)} e^{2\pi i(jr/m+ks/n)} e^{i(j\xi+k\eta)} \right)$$

$$= u^* W_{mn}^* F_{mn}^* D_{mn} F_{mn} W_{mn} u. \tag{6.8}$$

Here W_{mn} is a diagonal matrix defined by

$$W_{mn} \equiv W_m^\xi \otimes W_n^\eta$$

with

$$W_m^\xi = \text{diag}\left(1, e^{i\xi}, \cdots, e^{i(m-1)\xi}\right)$$

and

$$W_n^\eta = \text{diag}\left(1, e^{i\eta}, \cdots, e^{i(n-1)\eta}\right);$$

D_{mn} is a diagonal matrix defined by

$$D_{mn} \equiv \text{diag}(D_{(0)}, D_{(1)}, \cdots, D_{(m-1)})$$

with

$$D_{(i)} = \text{diag}\left((f(x_i, y_0), (f(x_i, y_1), \cdots, (f(x_i, y_{n-1}))\right)$$

for $i = 0, 1, \cdots, m-1$; and

$$F_{mn} \equiv F_m \otimes F_n$$

with the Fourier matrices F_m and F_n. Let

$$A_{mn}(f) \equiv W_{mn}^* F_{mn}^* D_{mn} F_{mn} W_{mn}. \tag{6.9}$$

The $A_{mn}(f)$ is called the block $\{\omega\}$-circulant preconditioner. By using (6.7) and (6.9), we note that $A_{mn}(f)$ is Hermitian positive definite. Let $v = A_{mn}(f)^{1/2}u$, we get

$$v^* A_{mn}(f)^{-1/2} T_{mn}(f) A_{mn}(f)^{-1/2} v \approx v^* v.$$

By using FFTs, the matrix-vector multiplication

$$A_{mn}(f)^{-1}v = W_{mn}^* F_{mn}^* D_{mn}^{-1} F_{mn} W_{mn} v$$

requires $O(mn \log mn)$ operations. Therefore, $A_{mn}(f)$ seems to be a good preconditioner of $T_{mn}(f)$. We will prove that the eigenvalues of $A_{mn}(f)^{-1} T_{mn}(f)$ are clustered around 1 except $O(m) + O(n)$ outliers.

6.3   Clustering of eigenvalues

Now, we consider the approximation error. By using (6.8) and (6.9) we have

$$u^* T_{mn}(f)u = \sum_{j,k=0}^{m-1} \sum_{p,q=0}^{n-1} \bar{u}_p^{(j)} t_{j-k,p-q} u_q^{(k)}$$

$$\approx \sum_{j,k=0}^{m-1} \sum_{p,q=0}^{n-1} \bar{u}_p^{(j)} \tilde{t}_{j-k,p-q} u_q^{(k)}$$

$$= u^* A_{mn}(f)u$$

where

$$\tilde{t}_{j,k} = \tilde{t}_{j,k}(f) \equiv \frac{1}{mn} \sum_{r=0}^{m-1} \sum_{s=0}^{n-1} f(x_r, y_s) e^{-2\pi i(jr/m + ks/n)} e^{-i(j\xi + k\eta)} \tag{6.10}$$

for $j, k = 0, \pm 1, \pm 2, \cdots$. The approximation error depends on $t_{j,k}$ and $\tilde{t}_{j,k}$. Now replacing $f(x_r, y_s)$ by the Fourier series of f at (x_r, y_s) in (6.10), we obtain

$$\tilde{t}_{j,k} = \frac{1}{mn} \sum_{r=0}^{m-1} \sum_{s=0}^{n-1} \left(\sum_{p,q \in \mathbb{Z}} t_{p,q} e^{i(px_r + qy_s)} \right) e^{-2\pi i(jr/m + ks/n)} e^{-i(j\xi + k\eta)}$$

$$= t_{j,k} + \sum_{\alpha,\beta \in \mathbb{Z} \backslash \{0\}} t_{j+\alpha m, k+\beta n} e^{i(\alpha m \xi + \beta n \eta)}$$

$$+ \sum_{\alpha \in \mathbb{Z} \backslash \{0\}} t_{j+\alpha m,k} e^{i\alpha m \xi} + \sum_{\beta \in \mathbb{Z} \backslash \{0\}} t_{j,k+\beta n} e^{i\beta n \eta} \qquad (6.11)$$

where $\mathbb{Z}$ is the set of all integers.

Let

$$b_{j,k} = b_{j,k}(f) \equiv \sum_{\alpha,\beta \in \mathbb{Z} \backslash \{0\}} t_{j+\alpha m, k+\beta n} e^{i(\alpha m \xi + \beta n \eta)}$$

$$+ \sum_{\alpha \in \mathbb{Z} \backslash \{0\}} t_{j+\alpha m,k} e^{i\alpha m \xi} + \sum_{\beta \in \mathbb{Z} \backslash \{0\}} t_{j,k+\beta n} e^{i\beta n \eta}, \qquad (6.12)$$

then it follows by using (6.11) and (6.12) that

$$T_{mn}(f) = A_{mn}(f) - B_{mn}(f). \qquad (6.13)$$

Here $B_{mn}(f)$ is a BTTB matrix defined by $b_{j,k}$, i.e., the (r,s)-th entry of the (p,q)-th block of $B_{mn}(f)$ is given by $b_{p-q,r-s}$. Thus

$$A_{mn}(f)^{-1} T_{mn}(f) = I - A_{mn}(f)^{-1} B_{mn}(f). \qquad (6.14)$$

Lemma 6.1 *Let $p_{s,t} \geq 0$ be a trigonometric polynomial of the form*

$$p_{s,t}(x,y) \equiv \sum_{p=-s}^{s} \sum_{q=-t}^{t} t_{p,q} e^{i(px+qy)}$$

where $t_{p,q}$ are constants. If $2s \leq m$ and $2t \leq n$, then there are at most $O(m) + O(n)$ eigenvalues of $A_{mn}(p_{s,t})^{-1} T_{mn}(p_{s,t})$ which are different from 1.

Proof: By using the definition of $t_{j,k}$, it follows that

$$t_{j+\alpha m, k+\beta n}(p_{s,t}) = 0$$

for $|j| \leq m - 1 - s$ or $|k| \leq n - 1 - t$;

$$t_{j+\alpha m, k}(p_{s,t}) = 0$$

for $|j| \leq m - 1 - s$ or $|k| \geq t + 1$; and

$$t_{j, k+\beta n}(p_{s,t}) = 0$$

for $|j| \geq s + 1$ or $|k| \leq n - 1 - t$. By using (6.12) we have

$$\text{rank}((B_{mn}(p_{s,t})) \leq O(m) + O(n).$$

Now the result follows from (6.14). $\square$

In order to prove our main theorem, we need the following lemma.

Lemma 6.2 *Let $g \in \mathcal{C}_{2\pi \times 2\pi}$ be a non-negative function with finite number of zeros and $h \in \mathcal{C}_{2\pi \times 2\pi}$ be a positive function with $h_{\min} > 0$. Define $f \equiv gh$. Then, the eigenvalues of $T_{mn}(g)^{-1}T_{mn}(f)$ lie in the interval $[h_{\min}, h_{\max}]$.*

Proof: By the Mean Value Theorem of Integration, we have for any $u \in \mathbb{C}^{mn}$,

$$u^* T_{mn}(f)u$$

$$= \frac{1}{4\pi^2} \int_{-\pi}^{\pi}\int_{-\pi}^{\pi} f(x,y) \left| \sum_{j=0}^{m-1}\sum_{k=0}^{n-1} u_k^{(j)} e^{-i(jx+ky)} \right|^2 dx dy$$

$$= \frac{1}{4\pi^2} \int_{-\pi}^{\pi}\int_{-\pi}^{\pi} g(x,y)h(x,y) \left| \sum_{j=0}^{m-1}\sum_{k=0}^{n-1} u_k^{(j)} e^{-i(jx+ky)} \right|^2 dx dy$$

$$= h_* \frac{1}{4\pi^2} \int_{-\pi}^{\pi}\int_{-\pi}^{\pi} g(x,y) \left| \sum_{j=0}^{m-1}\sum_{k=0}^{n-1} u_k^{(j)} e^{-i(jx+ky)} \right|^2 dx dy$$

$$= h_* u^* T_{mn}(g)u$$

where $h_* \in [h_{\min}, h_{\max}]$. Since the matrix $T_{mn}(g)$ is positive definite by Theorem 6.1, we have

$$h_{\min} \leq h_* = \frac{u^* T_{mn}(f) u}{u^* T_{mn}(g) u} \leq h_{\max}$$

for any non-zero vector $u \in \mathbb{C}^{mn}$. By Courant–Fischer's Minimax Theorem, the eigenvalues of $T_{mn}(g)^{-1} T_{mn}(f)$ lie in the interval $[h_{\min}, h_{\max}]$.
□

The following definition is proposed in [69].

Definition 6.1 *Let $f(x_0, y_0) = 0$. The function $F_{(x_0,y_0)}(\tau)$ is defined by*

$$F_{(x_0,y_0)}(\tau) \equiv f(x_0 + \tau(x - x_0), y_0 + \tau(y - y_0)), \qquad \forall (x, y) \neq (x_0, y_0).$$

Then the zero (x_0, y_0) is said to be of order N if N is the smallest positive integer such that the $(N+1)$-th derivative $F_{(x_0,y_0)}^{(N+1)}(\tau)$ is continuous in a neighborhood of (x_0, y_0) and $F_{(x_0,y_0)}^{(N)}(\tau) \neq 0$ for all $(x, y) \neq (x_0, y_0)$.

It is proved in [69] that if $f \geq 0$ with zero (x_0, y_0) of order N, then $F_{(x_0,y_0)}^{(N)}(0)$ is positive for all $(x, y) \neq (x_0, y_0)$ and therefore N must be even. In the following, for simplicity, we restrict our attention to non-negative generating functions $f \in C_{2\pi \times 2\pi}$ with only one zero attained at $(0,0)$ of order 2ν. Let

$$p(x, y) = (2 - \cos x - \cos y)^\nu, \tag{6.15}$$

then p has zero attained at $(0,0)$ of order 2ν. Moreover, this trigonometric polynomial matches the zero of f, i.e., f/p is continuous and greater than zero on $[-\pi, \pi] \times [-\pi, \pi]$, see [69]. Now, we are ready to prove the main theorem in this chapter.

Theorem 6.3 *Let $f \in C_{2\pi \times 2\pi}$ be a non-negative function having zero at $(0,0)$ of order 2ν. Then the eigenvalues of $A_{mn}(f)^{-1} T_{mn}(f)$ are clustered around 1 except at most $O(m) + O(n)$ outliers.*

Proof: Let $h = f/p$ where p is defined as in (6.15). Then $h(x, y) > 0$ and in $C_{2\pi \times 2\pi}$. Moreover, for any $\epsilon > 0$, by Stone–Weierstrass Theorem, see [72], there exists a positive trigonometric polynomial

$$g(x, y) = \sum_{p=-s}^{s} \sum_{q=-t}^{t} t_{p,q} e^{i(px+qy)}$$

with $t_{-p,-q} = \bar{t}_{p,q}$ such that

$$|||h - g|||_\infty \leq \frac{1}{2}\epsilon h_{\min},$$

i.e.,

$$g(x,y) - \frac{1}{2}\epsilon h_{\min} \leq h(x,y) \leq g(x,y) + \frac{1}{2}\epsilon h_{\min} \qquad (6.16)$$

for all $(x,y) \in [-\pi, \pi] \times [-\pi, \pi]$. Since $p \geq 0$, we have

$$gp - \frac{1}{2}\epsilon h_{\min}p \leq f \leq gp + \frac{1}{2}\epsilon h_{\min}p. \qquad (6.17)$$

From the right-hand inequality in (6.17), we obtain by using (6.6),

$$u^*T_{mn}(f)u \leq u^*T_{mn}(gp)u + \frac{1}{2}\epsilon h_{\min}u^*T_{mn}(p)u$$

where $u \in \mathbb{C}^{mn}$. Furthermore, since $A_{mn}(f)$ is Hermitian positive definite, we have

$$\frac{u^*T_{mn}(f)u}{u^*A_{mn}(f)u} \leq \frac{u^*T_{mn}(gp)u}{u^*A_{mn}(f)u} + \frac{1}{2}\epsilon h_{\min}\frac{u^*T_{mn}(p)u}{u^*A_{mn}(f)u} \qquad (6.18)$$

for all non-zero $u \in \mathbb{C}^{mn}$. Now, it holds by (6.13) and Lemma 6.1 that

$$T_{mn}(p) = A_{mn}(p) + \tilde{R}_{mn} \qquad (6.19)$$

and

$$T_{mn}(gp) = A_{mn}(gp) + \hat{R}_{mn} \qquad (6.20)$$

where

$$\text{rank}(\tilde{R}_{mn}) \leq O(m) + O(n)$$

and

$$\text{rank}(\hat{R}_{mn}) \leq O(m) + O(n).$$

Substituting (6.19) and (6.20) into (6.18), we obtain

$$\frac{u^*T_{mn}(f)u}{u^*A_{mn}(f)u} \leq \frac{u^*A_{mn}(gp)u}{u^*A_{mn}(f)u} + \frac{u^*\hat{R}_{mn}u}{u^*A_{mn}(f)u}$$

$$+ \frac{1}{2}\epsilon h_{\min}\frac{u^*A_{mn}(p)u}{u^*A_{mn}(f)u} + \frac{1}{2}\epsilon h_{\min}\frac{u^*\tilde{R}_{mn}u}{u^*A_{mn}(f)u}.$$

Since

$$\frac{u^* A_{mn}(p)u}{u^* A_{mn}(f)u} = \frac{1}{u^* A_{mn}(h)u} \leq \frac{1}{h_{\min}},$$

we get

$$\frac{u^*(T_{mn}(f) - R_{mn})u}{u^* A_{mn}(f)u} \leq \frac{u^* A_{mn}(gp)u}{u^* A_{mn}(f)u} + \frac{1}{2}\epsilon$$

where

$$R_{mn} = \frac{1}{2}\epsilon h_{\min} \tilde{R}_{mn} + \hat{R}_{mn}$$

with

$$\text{rank}(R_{mn}) \leq O(m) + O(n).$$

Now, by using the facts

$$A_{mn}(gp) = A_{mn}(g)A_{mn}(p) \quad \text{and} \quad A_{mn}(f) = A_{mn}(h)A_{mn}(p),$$

we have

$$\frac{u^*(T_{mn}(f) - R_{mn})u}{u^* A_{mn}(f)u} \leq \frac{u^* A_{mn}(g)u}{u^* A_{mn}(h)u} + \frac{1}{2}\epsilon. \tag{6.21}$$

Let $v = A_{mn}(p)^{1/2}u$ for all non-zero $u \in \mathbb{C}^{mn}$. By using (6.16) we have

$$v^* A_{mn}(g)v \leq v^* A_{mn}(h)v + \frac{1}{2}\epsilon h_{\min}v^*v.$$

Since, by using Lemma 6.2,

$$0 < \frac{v^*v}{v^* A_{mn}(h)v} = \frac{u^* A_{mn}(p)u}{u^* A_{mn}(f)u} \leq \frac{1}{h_{\min}},$$

we further have

$$\frac{u^* A_{mn}(g)u}{u^* A_{mn}(h)u} \leq 1 + \frac{1}{2}\epsilon.$$

Using the above inequality in (6.21), we obtain

$$\frac{u^*(T_{mn}(f) - R_{mn})u}{u^* A_{mn}(f)u} \leq 1 + \epsilon.$$

Similarly we conclude from the left-hand inequality in (6.17) that

$$\frac{u^*(T_{mn}(f) - R_{mn})u}{u^* A_{mn}(f)u} \geq 1 - \epsilon.$$

Consequently, since

$$\text{rank}(R_{mn}) \leq O(m) + O(n),$$

we know that at most $O(m) + O(n)$ eigenvalues of $A_{mn}(f)^{-1}T_{mn}(f)$ are not contained in $[1 - \epsilon, 1 + \epsilon]$. □

Hence, if the PCG method is applied to solving the BTTB system (6.5), we can expect a fast convergence rate. For more detailed discussions on the convergence and operation cost of the method, we refer to [68].

6.4   Numerical results

We apply the PCG method with the block {ω}-circulant preconditioner for solving the BTTB system

$$T_{mn}(f)x = b$$

where $T_{mn}(f)$ is generated by a non-negative function f having a finite number of zeros. We test the following three systems with different generating functions defined on $[-\pi, \pi] \times [-\pi, \pi]$. They are

(i) $f_1(x,y) = x^2 + y^2$;

(ii) $f_2(x,y) = x^2 + y^4$;

(iii) $f_3(x,y) = (x^2 - 1)^2 y^2$.

The systems (i) and (ii) have a zero at $(0,0)$ and the system (iii) has zeros at $(1,0)$ and $(-1,0)$. In all tests, we used the vector of all ones as the right hand side vector b and the zero vector as the initial guess. The stopping criteria is

$$\frac{\|r^{(k)}\|_2}{\|r^{(0)}\|_2} < 10^{-7}$$

where $r^{(k)}$ is the residual vector after k-th iterations. Tables 6.1 to 6.3 show the number of iterations required for convergence with different choices of preconditioners. In these tables, I denotes no preconditioner and $A_{mn}(f_i)(i = 1, 2, 3)$ denotes block {ω}-circulant preconditioner. Iteration numbers greater than 10000 are denoted by '⋆'. For a comparison, the BCCB preconditioner $c_{F,F}^{(2)}(T_{mn})$ proposed in Chapter 2 is also tested. From numerical results, there is evidence that the performance of the block $\{\omega\}$-circulant preconditioner is better than that of the preconditioner $c_{F,F}^{(2)}(T_{mn})$ for ill conditioned problems. We would like to remark that if $f(x,y) = g(x)h(y)$, given as in the system (iii), then the corresponding matrix is a block Toeplitz matrix with tensor structure. We could use the fast algorithm proposed in Chapter 4 to obtain a faster convergence rate.

Table 6.1. Number of iterations for $f_1(x,y) = x^2 + y^2$

$m = n$	n^2	I	$c_{F,F}^{(2)}(T_{mn}(f_1))$	$A_{mn}(f_1)$
8	64	10	10	7
16	256	32	14	11
32	1024	75	20	11
64	4096	161	29	13
128	16384	333	46	16
256	65536	681	73	16

Table 6.2. Number of iterations for $f_2(x,y) = x^2 + y^4$

$m = n$	n^2	I	$c_{F,F}^{(2)}(T_{mn}(f_2))$	$A_{mn}(f_2)$
8	64	19	14	12
16	256	95	28	16
32	1024	291	56	26
64	4096	781	122	37
128	16384	2032	267	60
256	65536	4958	621	101

Table 6.3. Number of iterations for $f_3(x,y) = (x^2 - 1)^2 y^2$

$m = n$	n^2	I	$c_{F,F}^{(2)}(T_{mn}(f_3))$	$A_{mn}(f_3)$
8	64	37	18	21
16	256	359	64	50
32	1024	2608	125	34
64	4096	$\star$	271	45
128	16384	$\star$	559	73
256	65536	$\star$	1260	71

Chapter 7

Non-Circulant Block Preconditioners

In this chapter we study non-circulant block preconditioners. All of these preconditioners have been proved to be good and useful preconditioners for solving some block Toeplitz systems from practical problems, see [8, 18, 25, 27, 4, 52, 77].

7.1   Block band Toeplitz preconditioners

Band Toeplitz matrices are proposed in [27] as preconditioners for solving ill conditioned Toeplitz systems by the PCG method, see also Section 1.3.2. In this section, we extend the results in [27] to block Toeplitz systems. Again, let $T_{mn}(f)$ be assumed to be generated by a function $f(x, y) \in C_{2\pi \times 2\pi}$ with diagonals given by

$$t_{j,k}(f) \equiv t_k^{(j)}(f) = \frac{1}{4\pi^2} \int_{-\pi}^{\pi} \int_{-\pi}^{\pi} f(x, y) e^{-i(jx + ky)} dx dy,$$

for $j, k = 0, \pm 1, \pm 2, \cdots$. We then construct block band Toeplitz matrices B_{mn} with Toeplitz band blocks as our preconditioners to solve $T_{mn}u = b$ by the PCG method. The generating function $g(x, y)$ of $B_{mn}(g)$ is a trigonometric polynomial of fixed degree and is determined by minimizing the supremum norm

$$\left\vert\left\vert\left\vert \frac{f - g}{f} \right\vert\right\vert\right\vert_\infty.$$

7.1.1 Convergence analysis

For the block band Toeplitz preconditioner $B_{mn}(g)$, the following theorem gives a bound on the condition number of the preconditioned matrix $(B_{mn}(g))^{-1}T_{mn}(f)$.

Theorem 7.1 *Let $f \in C_{2\pi \times 2\pi}$ be the generating function of $T_{mn}(f)$ with $f \geq 0$ and g be the generating function of a block band Toeplitz matrix $B_{mn}(g)$:*

$$g(x,y) = \sum_{j=-M}^{M} \sum_{k=-N}^{N} b_{j,k} e^{i(jx+ky)}$$

with $b_{-j,-k} = \bar{b}_{j,k}$. If

$$\left\| \frac{f-g}{f} \right\|_{\infty} = h < 1,$$

then B_{mn} is positive definite and

$$\kappa(B_{mn}^{-1}T_{mn}) \leq \frac{1+h}{1-h}$$

for all $m, n > 0$.

Proof: By assumption we have

$$f(x,y)(1-h) \leq g(x,y) \leq f(x,y)(1+h),$$

for any $(x,y) \in [-\pi, \pi] \times [-\pi, \pi]$. It is clear that $g(x,y) \geq 0$. By Theorem 6.1, B_{mn} is positive definite for all $m, n > 0$. Since both T_{mn} and B_{mn} are BTTB matrices, we have

$$u^* T_{mn} u = \frac{1}{4\pi^2} \int_{-\pi}^{\pi} \int_{-\pi}^{\pi} \left| \sum_{j=0}^{m-1} \sum_{k=0}^{n-1} u_k^{(j)} e^{-i(jx+ky)} \right|^2 f(x,y) \, dx \, dy$$

and

$$u^* B_{mn} u = \frac{1}{4\pi^2} \int_{-\pi}^{\pi} \int_{-\pi}^{\pi} \left| \sum_{j=0}^{m-1} \sum_{k=0}^{n-1} u_k^{(j)} e^{-i(jx+ky)} \right|^2 g(x,y) \, dx \, dy$$

where $u \in \mathbb{C}^{mn}$ is given by

$$u = (u_0^{(0)}, u_1^{(0)}, \cdots, u_{n-1}^{(0)}, u_0^{(1)}, \cdots, u_0^{(m-1)}, \cdots, u_{n-1}^{(m-1)})^T.$$

Hence, we get

$$(1 - h)u^*T_{mn}u \leq u^*B_{mn}u \leq (1 + h)u^*T_{mn}u,$$

i.e.,

$$\frac{1}{1+h} \leq \frac{u^*T_{mn}u}{u^*B_{mn}u} \leq \frac{1}{1-h}.$$

By Courant–Fischer's Minimax Theorem, we know that

$$\frac{1}{1+h} \leq \lambda_{\min}(B_{mn}^{-1}T_{mn}) \leq \lambda_{\max}(B_{mn}^{-1}T_{mn}) \leq \frac{1}{1-h}.$$

We then have

$$\kappa(B_{mn}^{-1}T_{mn}) \leq \frac{1+h}{1-h}. \quad \square$$

By standard error analysis of the conjugate gradient method, see [38, 75], we know that the number of iterations for convergence is bounded by

$$\frac{1}{2}\left(\frac{1+h}{1-h}\right)\log\left(\frac{1}{\tau}\right) + 1$$

with tolerance τ. Since h can be found explicitly in Remez algorithm, we have a priori bound on the number of iterations for convergence.

7.1.2 Construction of preconditioner

We construct our block band Toeplitz preconditioner $B_{mn}(g)$ by Remez algorithm. Let the generating function g be given by

$$g(x,y) = \sum_{j=-M}^{M}\sum_{k=-N}^{N} b_{j,k}e^{i(jx+ky)}$$

with $b_{-j,-k} = \bar{b}_{j,k}$. Since the entries of $B_{mn}(g)$ are given by

$$(B_{mn}(g))_{p,q;s,t} = b_{p-q,s-t}(g),$$

$B_{mn}(g)$ is a band block Toeplitz matrix with Toeplitz band blocks.

In the case of $f(x,y) > 0$ on $[-\pi,\pi]\times[-\pi,\pi]$, we consider the following standard linear minimax approximation problem:

$$\text{minimize}_{p_{-M,-N},\ldots,p_{0,0},\ldots,p_{M,N}} |||1 - P(x,y)|||_\infty$$

where

$$P(x,y) = \sum_{j=-M}^{M} \sum_{k=-N}^{N} p_{j,k} \phi_{j,k}(x,y) \qquad (7.1)$$

with

$$p_{-j,-k} = \bar{p}_{j,k}, \quad \phi_{0,0} = 1/f(x,y) \quad \text{and} \quad \phi_{j,k}(x,y) = e^{i(jx+ky)}/f(x,y).$$

We use the optimal P to define our g by $P = g/f$. This optimal P (and hence g) can be obtained by a version of Remez algorithm proposed in [57] which can handle the case when $f(x_0, y_0) = 0$ for some $(x_0, y_0) \in [-\pi, \pi] \times [-\pi, \pi]$. We now describe this algorithm.

Given $\phi_{0,0}(x,y) = 1/f(x,y)$ and $\phi_{j,k}(x,y) = e^{i(jx+ky)}/f(x,y)$, we solve

Problem $\mathcal{P}$:

$$\text{Minimize } h$$

subject to

$$h \geq s \left(1 - \sum_{j=-M}^{M} \sum_{k=-N}^{N} p_{j,k} \phi_{j,k}(x,y) \right)$$

for

$$(s, x, y) \in \{-1, 1\} \times [-\pi, \pi] \times [-\pi, \pi].$$

One can think of Problem $\mathcal{P}$ as a linear programming problem (by replacing $[-\pi, \pi] \times [-\pi, \pi]$ by a finite set of points). The dual of this problem is given as follows.

Problem $\mathcal{D}$:

$$\text{Maximize } \sum_{s,x,y} s \cdot r_{s,x,y}$$

subject to

$$r_{s,x,y} \geq 0 \quad \text{and} \quad \sum_{s,x,y} r_{s,x,y} \phi_{j,k}(x,y) s = 0, \quad \forall j, k.$$

Thus, the simplex algorithm could be applied to solving Problem $\mathcal{D}$.

Suppose now that

$$f(x_0, y_0) = 0$$

for some $(x_0, y_0) \in [-\pi, \pi] \times [-\pi, \pi]$. Then by imposing the constraint $g(x_0, y_0) = 0$, i.e.,

$$\sum_{j=-M}^{M} \sum_{k=-N}^{N} p_{j,k} e^{i(jx_0 + ky_0)} = 0,$$

into Problem $\mathcal{P}$, we have the following Problem $\mathcal{P}'$.

Problem $\mathcal{P}'$:

$$\text{Minimize } h$$

subject to

$$h \geq s \left(1 - \sum_{j=-M}^{M} \sum_{k=-N}^{N} p_{j,k} \phi_{j,k}(x, y) \right)$$

for

$$(s, x, y) \in \{-1, 1\} \times [-\pi, \pi] \times [-\pi, \pi],$$

and

$$\sum_{j=-M}^{M} \sum_{k=-N}^{N} p_{j,k} e^{i(jx_0 + ky_0)} = 0. \tag{7.2}$$

The linear constraint (7.2) on $p_{j,k}$'s can be translated to its dual form in Problem $\mathcal{D}$, we then have Problem $\mathcal{D}'$ and we still can solve Problem $\mathcal{D}'$ by the simplex algorithm.

If $f \in C_{2\pi \times 2\pi}^3$, because $f \geq 0$, we have

$$f_x'(x_0, y_0) = f_y'(x_0, y_0) = 0.$$

Let the Hessian matrix

$$H(f)_{(x_0, y_0)} > 0.$$

Then in Problem $\mathcal{P}'$, if the Hessian matrix $H(g)_{(x_0, y_0)} > 0$ (which usually can be guaranteed by solving Problem $\mathcal{P}'$), we can also find the upper and low bounds of $P(x, y)$ defined by (7.1) in the following way. Because $g = P \cdot f$ is non-negative, we have

$$g_x'(x_0, y_0) = g_y'(x_0, y_0) = 0.$$

Furthermore, we have

$$P(x, y) = \frac{g(x, y)}{f(x, y)} = \frac{v^* H(g)_{(x_0, y_0)} v + o(\|v\|_2^2)}{v^* H(f)_{(x_0, y_0)} v + o(\|v\|_2^2)} \tag{7.3}$$

where $v^T = (x - x_0, y - y_0)$. From (7.3) we obtain

$$\frac{\lambda_{\min}(H(g)_{(x_0,y_0)})}{\lambda_{\max}(H(f)_{(x_0,y_0)})} \leq \liminf_{\|v\|_2 \to 0} P(x,y) \leq \limsup_{\|v\|_2 \to 0} P(x,y)$$

$$\leq \frac{\lambda_{\max}(H(g)_{(x_0,y_0)})}{\lambda_{\min}(H(f)_{(x_0,y_0)})}.$$

Then we have the following inequality

$$0 < r \leq P(x,y) = \frac{g(x,y)}{f(x,y)} \leq R < +\infty, \qquad (7.4)$$

for $(x,y) \in [-\pi, \pi] \times [-\pi, \pi]$, where r and R are two constants given by

$$r \equiv \frac{\lambda_{\min}(H(g)_{(x_0,y_0)})}{\lambda_{\max}(H(f)_{(x_0,y_0)})} \quad \text{and} \quad R \equiv \frac{\lambda_{\max}(H(g)_{(x_0,y_0)})}{\lambda_{\min}(H(f)_{(x_0,y_0)})}.$$

We remark that r and R in (7.4) can also give an upper bound of the condition number of the preconditioned matrix $B_{mn}^{-1} T_{mn}$.

Let us discuss the computational cost of our method. The complexity of solving Problem $\mathcal{D}$ (and hence Problem $\mathcal{P}$) by the simplex algorithm is $O(M^3 N^3)$, see [57]. We stress the fact that the construction of g is independent of mn. Thus, as long as f is fixed and Problem $\mathcal{D}$ (and hence Problem $\mathcal{P}$) is solved, the entries of B_{mn} are determined for all m, n. In each iteration of the PCG method, we have to compute matrix-vector multiplications $T_{mn} v$ and $B_{mn}^{-1} y$. We have already known that $T_{mn} v$ can be computed in $O(mn \log mn)$ operations, see Section 2.2.2. The vector $z = B_{mn}^{-1} y$ can be obtained by solving $B_{mn} z = y$ with some fast band solvers, see [36, 77]. Only $O(mn \log mn)$ operations are required by using those mothods. Since the number of iterations is independent of the size of the matrix, we conclude that the total complexity of our method remains $O(mn \log mn)$. For numerical results of the PCG method with block band Toeplitz preconditioners, we refer to [69].

7.2 Preconditioners based on fast transforms

Let

$$\mathcal{M}_{P \otimes Q} = \{(P^* \otimes Q^*)\Lambda_{mn}(P \otimes Q) \mid \Lambda_{mn} \text{ is a diagonal matrix}\}$$

where P is any given $m \times m$ unitary matrix and Q is any given $n \times n$ unitary matrix. The preconditioner M_{mn} is defined to be the minimizer of $\|T_{mn} - B_{mn}\|_F$ over all $B_{mn} \in \mathcal{M}_{P \otimes Q}$. Let Φ^s, Φ^c and Φ^h be defined as in Section 1.3.1. By using the same trick as in Section 2.3, one can prove the following lemma easily, see [4, 55].

Lemma 7.1 *Let f be a real-valued even function in the Wiener class. Then for all $\epsilon > 0$ there exist M, $N > 0$ such that for all $m > M$ and $n > N$, at most $O(m) + O(n)$ eigenvalues of $T_{mn}(f) - M_{mn}(f)$ have absolute values larger than ϵ. Here $M_{mn}(f) \in \mathcal{M}_{P \otimes Q}$, for $P \otimes Q = \Phi^s_m \otimes \Phi^s_n$, $\Phi^c_m \otimes \Phi^c_n$ and $\Phi^h_m \otimes \Phi^h_n$, respectively.*

The results in Lemma 7.1 could be extended from the Wiener class of functions to $C_{2\pi \times 2\pi}$ by using the same trick as we did in the proof of Theorem 3.1.

Theorem 7.2 *Let f be an even function in $C_{2\pi \times 2\pi}$. Then for all $\epsilon > 0$ there exist M, $N > 0$ such that for all $m > M$ and $n > N$, at most $O(m) + O(n)$ eigenvalues of $T_{mn}(f) - M_{mn}(f)$ have absolute values larger than ϵ. Here $M_{mn}(f) \in \mathcal{M}_{P \otimes Q}$, for $P \otimes Q = \Phi^s_m \otimes \Phi^s_n$, $\Phi^c_m \otimes \Phi^c_n$ and $\Phi^h_m \otimes \Phi^h_n$, respectively.*

By noting that

$$(M_{mn}(f))^{-1} T_{mn}(f) - I_{mn} = (M_{mn}(f))^{-1}(T_{mn}(f) - M_{mn}(f))$$

and Theorem 2.3 (ii), we have the following corollary.

Corollary 7.1 *Let f be an even function in $C_{2\pi \times 2\pi}$ and $f_{\min} > 0$. Then for all $\epsilon > 0$, there exist M, $N > 0$, such that for all $m > M$ and $n > N$, at most $O(m) + O(n)$ eigenvalues of the matrix $(M_{mn}(f))^{-1} T_{mn}(f) - I_{mn}$ have absolute values larger than ϵ. Here $M_{mn}(f) \in \mathcal{M}_{P \otimes Q}$, for $P \otimes Q = \Phi^s_m \otimes \Phi^s_n$, $\Phi^c_m \otimes \Phi^c_n$ and $\Phi^h_m \otimes \Phi^h_n$, respectively.*

From Theorem 6.1 and Corollary 7.1, we note that the number of iterations to achieve a fixed accuracy remains bounded as m, n are increased. Since the spectrum of the preconditioned matrix is clustered, we can expect a fast convergence rate when the PCG method is applied to solving the preconditioned system. In each iteration of the PCG method, we have to compute matrix-vector multiplication $T_{mn}v$ and solve the system $M_{mn}y = u$. The $T_{mn}v$ can be computed in $O(mn \log mn)$ operations, see Section 2.2.2. The system $M_{mn}y = u$ can also be solved in $O(mn \log mn)$ operations by using the 2-dimensional fast sine transform for $M_{mn} \in \mathcal{M}_{\Phi_m^s \otimes \Phi_n^s}$; by using the 2-dimensional fast cosine transform for $M_{mn} \in \mathcal{M}_{\Phi_m^c \otimes \Phi_n^c}$; or by using the 2-dimensional fast Hartley transform for $M_{mn} \in \mathcal{M}_{\Phi_m^h \otimes \Phi_n^h}$.

Chapter 8

Multigrid Block Toeplitz Solvers

We study the solutions of BTTB systems $T_{mn}u = b$ by multigrid methods (MGMs). Here BTTB matrices T_{mn} are assumed to be generated by a non-negative function $f(x,y)$ with zeros. Since by Theorems 6.1 and 6.2, the matrices T_{mn} are ill conditioned, the convergence factor of classical stationary iterative methods will approach 1 as the size of matrices becomes large. These classical methods, therefore, are not applicable. The MGM is then proposed in this chapter.

8.1 Introduction

Let $T_{mn}u = b$ be a BTTB system with diagonals $t_{j,k}$ of T_{mn} generated by an even function $f(x,y) \in \mathcal{C}_{2\pi \times 2\pi}$ as follows,

$$t_{j,k}(f) \equiv t_k^{(j)}(f) = \frac{1}{4\pi^2} \int_{-\pi}^{\pi} \int_{-\pi}^{\pi} f(x,y)e^{-i(jx+ky)} dx dy,$$

for $j, k = 0, \pm 1, \pm 2, \cdots$. By Theorems 6.1 and 6.2, we know that when $f(x,y)$ is non-negative and vanishes at some points in $[-\pi, \pi] \times [-\pi, \pi]$, then the condition number $\kappa(T_{mn})$ of T_{mn} is unbounded as m, n tend to infinity, i.e., T_{mn} is ill conditioned. The convergence factor of classical stationary iterative methods for solving such kind of system is expected to approach 1 for large m and n. Thus, the Jacobi, Gauss–Seidel or SOR methods are not applicable.

Besides the PCG method with preconditioners proposed in Chapters 6 and 7, we consider the use of the MGM as an alternative way for solving ill conditioned BTTB systems. First of all, let us introduce the MGM scheme. For given BTTB systems

$$T_{mn}u = b$$

with $u \in \mathbb{R}^{mn}$, we define a sequence of sub-systems on different levels:

$$T^j u^j = b^j$$

where $u^j \in \mathbb{R}^{m_j n_j}$, for $1 \leq j \leq q$, and q is the total number of levels with $j = 1$ being the finest level. Therefore $T^1 = T_{mn}$ and $m_1 n_1 = mn$. For $j > 1$, $m_j n_j$ are just the sizes of the matrices T^j. We denote the interpolation and restriction operators by

$$I_{j+1}^j : \mathbb{R}^{m_{j+1} n_{j+1}} \longrightarrow \mathbb{R}^{m_j n_j}$$

and

$$I_j^{j+1} : \mathbb{R}^{m_j n_j} \longrightarrow \mathbb{R}^{m_{j+1} n_{j+1}}$$

respectively. We will choose

$$I_j^{j+1} = (I_{j+1}^j)^T.$$

The coarse grid operators are defined by the Galerkin algorithm, i.e.,

$$T^{j+1} = I_j^{j+1} T^j I_{j+1}^j, \tag{8.1}$$

for $1 \leq j \leq q$. Thus, if T^j is symmetric positive definite, so is T^{j+1}. The smoothing operator is denoted by

$$S^j : \mathbb{R}^{m_j n_j} \longrightarrow \mathbb{R}^{m_j n_j}.$$

Typical smoothing operators are the Jacobi, Gauss–Seidel or SOR iterations, see for instance [12]. Once the components above are fixed, a multigrid cycling procedure can be set up.

We will concentrate on the MGM scheme with V-cycle which is given as follows, see [12].

$$
\begin{cases}
\text{Procedure } \mathbf{MGM}(\nu_1, \nu_2)(u^j, b^j); \\
\qquad \text{if } j = q \\
\qquad\qquad \text{then } u^q := (T^q)^{-1} b^q, \\
\qquad \text{end if} \\
\qquad \text{begin} \\
\qquad\qquad \text{for } i = 1, \cdots, \nu_1, \text{ do} \\
\qquad\qquad\qquad u^j := S^j u^j + (I_{m_j n_j} - S^j)(T^j)^{-1} b^j, \\
\qquad\qquad \text{end do} \\
\qquad\qquad d^{j+1} := I_j^{j+1}(T^j u^j - b^j), \\
\qquad\qquad e_0^{j+1} := 0, \\
\qquad\qquad e^{j+1} := \mathbf{MGM}(\nu_1, \nu_2)(e_0^{j+1}, d^{j+1}); \\
\qquad\qquad u^j := u^j - I_{j+1}^j e^{j+1}, \\
\qquad\qquad \text{for } i = 1, \cdots, \nu_2, \text{ do} \\
\qquad\qquad\qquad u^j := S^j u^j + (I_{m_j n_j} - S^j)(T^j)^{-1} b^j, \\
\qquad\qquad \text{end do} \\
\qquad\qquad \mathbf{MGM}(\nu_1, \nu_2) := u^j; \\
\qquad \text{end}
\end{cases}
$$

where $I_{m_j n_j}$ is the identity matrix. When $q = 2$, we have the two-grid method (TGM).

We show that for a class of ill conditioned BTTB matrices, the convergence factor of the TGM with our interpolation operator is uniformly bounded below 1 independent of m and n. We then prove that for this class of BTTB matrices the convergence factor of the MGM with V-cycle will be level-dependent. We remark however that our numerical results in Section 8.4 show that the MGM with V-cycle has already given a convergence rate of independent level. Since the cost per iteration is of $O(mn \log mn)$ operations, see Section 8.4, the total cost for solving the system is therefore of $O(mn \log mn)$ operations. For the history of the TGM and MGM Toeplitz solvers, we refer to [19, 32, 35, 36, 88, 89].

8.2 Convergence rate of TGM

In this section, we prove the convergence results of the TGM for BTTB systems with an even generating function $f(x, y) \in C_{2\pi \times 2\pi}$ satisfying

$$\min_{(x,y)\in[-\pi,\pi]^2} \frac{f(x,y)}{2 - \cos x - \cos y} > 0 \tag{8.2}$$

or

$$\min_{(x,y)\in[-\pi,\pi]^2} \frac{f(x,y)}{2 + \cos x + \cos y} > 0. \tag{8.3}$$

For these level-2 symmetric BTTB systems, the block restriction operator I_h^H in the TGM is defined by the following block projector,

$$I_h^H = P \otimes P \tag{8.4}$$

where

$$P = \begin{bmatrix} 1\,2\,1 & & \\ & 1 \ \ 2 \ \ 1 & \\ & & \ddots \ \ \ddots \ \ \ddots \end{bmatrix} \tag{8.5}$$

or

$$P = \begin{bmatrix} -1\,2\,-1 & & \\ & -1 \ \ 2 \ \ -1 & \\ & & \ddots \ \ \ddots \ \ \ddots \end{bmatrix} \tag{8.6}$$

depending on whether (8.2) or (8.3) holds. The block interpolation operator is defined as

$$I_H^h = (I_h^H)^T.$$

Let us introduce the following notations for convenience. We say $A > B$ (respectively $A \geq B$) if $A - B$ is a positive definite (respectively positive semi-definite) matrix. For $A > 0$, the following inner products are useful in the convergence analysis of the TGM and MGM, see [74]:

$$\langle u, v \rangle_{\bar{0}} = \langle \delta(A)u, v \rangle, \quad \langle u, v \rangle_{\bar{1}} = \langle Au, v \rangle,$$
$$\langle u, v \rangle_{\bar{2}} = \langle (\delta(A))^{-1}Au, Av \rangle. \tag{8.7}$$

Here $\langle \cdot, \cdot \rangle$ is the Euclidean inner product and $\delta(A)$, defined as in Section 1.2.2, is the diagonal matrix whose diagonal is equal to the diagonal of A. The respective norms of the inner products defined as in (8.7) are denoted by $\| \cdot \|_{\bar{i}}$ for $i = 0, 1, 2$. The damped–Jacobi smoother is defined by

$$S = I - \omega(\delta(A))^{-1}A, \tag{8.8}$$

see [12]. Then $\|S\|_{\bar{1}} \leq 1$ if ω is properly chosen. We have the following theorem, see [74].

Theorem 8.1 *Suppose that the BTTB matrix $T_{mn} > 0$. Let α be chosen such that*

$$\alpha \leq \frac{t_{0,0}}{\rho(T_{mn})} \tag{8.9}$$

where $t_{0,0}$ is the main diagonal entry of T_{mn} and $\rho(T_{mn})$ is the spectral radius of T_{mn}. Let S be the smoother defined as in (8.8) with $A = T_{mn}$ and $\omega = \alpha$, then

$$\|Se\|_{\tilde{1}}^2 \leq \|e\|_{\tilde{1}}^2 - \alpha\|e\|_{\tilde{2}}^2, \qquad \forall e \in \mathbb{R}^{mn}. \tag{8.10}$$

Inequality (8.10) is called the *smoothing condition*. We see from Theorem 8.1 that $\|S\|_{\tilde{1}} \leq 1$.

Let $T^h \equiv T_{mn}$ be the fine grid operator and let $T^H \equiv I_h^H T^h I_H^h$ be the coarse grid operator. For the TGM the correction operator is given by

$$C^h = I - I_H^h (T^H)^{-1} I_h^H T^h,$$

and the smoothing operator defined as in (8.8) is given by

$$S^h = I - \alpha(\delta(T^h))^{-1} T^h.$$

Usually, the global iteration matrix of TGM is given by $(S^h)^{\nu_2} C^h (S^h)^{\nu_1}$ with the convergence factor given by

$$\|(S^h)^{\nu_2} C^h (S^h)^{\nu_1}\|_{\tilde{1}},$$

see [74]. In the following we will use the global iteration matrix

$$G^h = S^h C^h$$

for simplicity. The following theorem can be found in [74].

Theorem 8.2 *Let $T^h \equiv T_{mn} > 0$ and α be chosen such that S^h satisfies the smoothing condition (8.10), i.e.,*

$$\|S^h e^h\|_{\tilde{1}}^2 \leq \|e^h\|_{\tilde{1}}^2 - \alpha\|e^h\|_{\tilde{2}}^2, \qquad \forall e^h \in \mathbb{R}^{mn},$$

where α is given by (8.9). Suppose that the projection operator I_h^H has full rank and that there exists a scalar $\beta > 0$ such that

$$\min_{e^H \in \mathbb{R}^{\lfloor mn/4 \rfloor}} \|e^h - I_H^h e^H\|_{\tilde{0}}^2 \leq \beta\|e^h\|_{\tilde{1}}^2, \qquad \forall e^h \in \mathbb{R}^{mn}. \tag{8.11}$$

Then $\beta \geq \alpha$ and the convergence factor of the h-H two-level TGM satisfies

$$\|G^h\|_{\tilde{1}} \leq \sqrt{1 - \frac{\alpha}{\beta}}.$$

The inequality (8.11) is called the *correcting condition*. From Theorems 8.1 and 8.2, if α is chosen as in (8.9), then we only have to establish (8.11) in order to get the convergence results. We have the following theorem.

Theorem 8.3 *Let T_{mn} be generated by an even function $f(x, y) \in \mathcal{C}_{2\pi \times 2\pi}$ satisfying (8.2) or (8.3). Let I_h^H be chosen as in (8.4) with P given by (8.5) or (8.6) accordingly. Then there exists a scalar $\beta > 0$ which is independent of m and n such that (8.11) holds. In particular, the convergence factor of the TGM is bounded uniformly below 1.*

Proof: We only prove the theorem for the case of (8.2). Let $\mathbb{N}$ denote the set of all natural numbers. We first assume that $m = 2k+1$ and $n = 2l+1$ for some $k, l \in \mathbb{N}$. For any

$$e^h = (e_{1,1}, e_{1,2}, \cdots, e_{1,n}, e_{2,1}, \cdots, e_{m,1}, \cdots, e_{m,n})^T \in \mathbb{R}^{mn}$$

we define

$$e^H = (\tilde{e}_{1,1}, \tilde{e}_{1,2}, \cdots, \tilde{e}_{1,l}, \tilde{e}_{2,1}, \cdots, \tilde{e}_{k,1}, \cdots, \tilde{e}_{k,l})^T \in \mathbb{R}^{kl}$$

where

$$\tilde{e}_{i,j} = (1/4)e_{2i,2j}, \qquad 1 \le i \le k, \quad 1 \le j \le l.$$

For simplicity we set $e_{i,j} = 0$ for $i \le 0$ or $j \le 0$ or $i > m$ or $j > n$. With I_h^H defined as in (8.4), P given by (8.5) and $\| \cdot \|_{\bar{0}}$ defined as in (8.7), we have

$$\|e^h - I_H^h e^H\|_{\bar{0}}^2$$

$$= t_{0,0} \sum_{i=0}^{k} \sum_{j=0}^{l} \Bigg\{ \left(e_{2i+1,2j+1} \right.$$

$$- \frac{1}{4} e_{2i,2j} - \frac{1}{4} e_{2i,2j+2} - \frac{1}{4} e_{2i+2,2j}$$

$$- \frac{1}{4} e_{2i+2,2j+2} \right)^2 + \left(e_{2i+1,2j} - \frac{1}{2} e_{2i,2j} - \frac{1}{2} e_{2i+2,2j} \right)^2$$

$$+ \left(e_{2i,2j+1} - \frac{1}{2} e_{2i,2j} - \frac{1}{2} e_{2i,2j+2} \right)^2 \Bigg\}. \qquad (8.12)$$

Thus (8.11) is proved if we can bound the right hand side above by $\beta \|e^h\|_{\bar{1}}$ for some β which is independent of e^h. For the first square term in the right hand side of (8.12), we have

$$\sum_{i=0}^{k}\sum_{j=0}^{l}(e_{2i+1,2j+1} - \frac{1}{4}e_{2i,2j} - \frac{1}{4}e_{2i,2j+2} - \frac{1}{4}e_{2i+2,2j} - \frac{1}{4}e_{2i+2,2j+2})^2$$

$$= \sum_{i=0}^{k}\sum_{j=0}^{l}\left\{ e_{2i+1,2j+1}^2 + \frac{1}{16}e_{2i,2j}^2 + \frac{1}{16}e_{2i,2j+2}^2 + \frac{1}{16}e_{2i+2,2j}^2 + \frac{1}{16}e_{2i+2,2j+2}^2 \right.$$

$$-\frac{1}{2}e_{2i+1,2j+1}(e_{2i,2j} + e_{2i,2j+2} + e_{2i+2,2j} + e_{2i+2,2j+2})$$

$$+\frac{1}{8}(e_{2i,2j}e_{2i,2j+2} + e_{2i,2j}e_{2i+2,2j} + e_{2i,2j}e_{2i+2,2j+2}$$

$$\left. +e_{2i,2j+2}e_{2i+2,2j} + e_{2i,2j+2}e_{2i+2,2j+2} + e_{2i+2,2j}e_{2i+2,2j+2}) \right\}$$

$$\leq \sum_{i=1}^{m}\sum_{j=1}^{n}\left\{ e_{i,j}^2 - \frac{1}{2}e_{i,j}(e_{i-1,j+1} + e_{i+1,j-1}) \right\}$$

$$= \langle e^h, T_{mn}(1 - \cos x \cos y)e^h \rangle \qquad (8.13)$$

where $T_{mn}(1 - \cos x \cos y)$ is the BTTB matrix generated by $1 - \cos x \cos y$. Since

$$1 - \cos x \cos y \leq 2 - \cos x - \cos y,$$

we have

$$\langle e^h, T_{mn}(1 - \cos x \cos y)e^h \rangle \leq \langle e^h, T_{mn}(2 - \cos x - \cos y)e^h \rangle. \qquad (8.14)$$

For the other square terms on (8.12), by following the proof of Theorem 3 in [88], one can easily find

$$\sum_{i=0}^{k}\sum_{j=0}^{l}\left\{ (e_{2i+1,2j} - \frac{1}{2}e_{2i,2j} - \frac{1}{2}e_{2i+2,2j})^2 \right.$$

$$\left. +(e_{2i,2j+1} - \frac{1}{2}e_{2i,2j} - \frac{1}{2}e_{2i,2j+2})^2 \right\}$$

$$\leq \langle e^h, (T_m(1 - \cos x) \otimes I_n)e^h \rangle + \langle e^h, (I_m \otimes T_n(1 - \cos y))e^h \rangle$$

$$= \langle e^h, T_{mn}(2 - \cos x - \cos y)e^h \rangle. \qquad (8.15)$$

Thus, by combining (8.13), (8.14) and (8.15), we have

$$\min_{e^H \in \mathbb{R}^{\lfloor kl/4 \rfloor}} \|e^h - I_H^h e^H\|_0^2 \leq 2t_{0,0}\langle e^h, T_{mn}(2 - \cos x - \cos y)e^h \rangle.$$

In order to establish (8.11) we only have to prove that

$$2t_{0,0}\langle e^h, T_{mn}(2 - \cos x - \cos y)e^h \rangle \leq \beta\|e^h\|_1^2, \qquad \forall e^h \in \mathbb{R}^{mn},$$

for some β which is independent of e^h. By definition of $\| \cdot \|_{\bar{1}}$, see (8.7), it is equivalent to proving

$$2t_{0,0}\langle e^h, T_{mn}(2 - \cos x - \cos y)e^h \rangle \le \beta \langle e^h, T^h e^h \rangle, \qquad \forall e^h \in \mathbb{R}^{mn},$$

for some β which is independent of e^h. By (8.2) and Theorem 2.4, it is obvious that

$$\gamma T_{mn}(2 - \cos x - \cos y) \le T_{mn}(f) = T_{mn} \equiv T^h$$

where

$$\gamma = \min_{(x,y)\in[-\pi,\pi]^2} \frac{f(x,y)}{2 - \cos x - \cos y} > 0.$$

Hence

$$2t_{0,0}\langle e^h, T_{mn}(2 - \cos x - \cos y)e^h \rangle \le \beta \langle e^h, T^h e^h \rangle, \qquad \forall e^h \in \mathbb{R}^{mn}, \quad (8.16)$$

where

$$\beta = \frac{2t_{0,0}}{\gamma}.$$

Thus (8.11) holds for the case of $m = 2k + 1$ and $n = 2l + 1$.

Next we consider the case where m is not of the form $2k + 1$ and n is not of the form $2l + 1$. In this case, we let $k = m/2$, $\tilde{m} = 2k + 1 > m$, and $l = n/2$, $\tilde{n} = 2l + 1 > n$. We then embed the vector e^h into longer vectors $\tilde{e}^h$ of size $\tilde{m}\tilde{n}$ by zeros. Since

$$\|e^h - I_H^h e^H\|_0^2 \le \|\tilde{e}^h - \tilde{I}_H^h \tilde{e}^H\|_0^2$$

and

$$\langle \tilde{e}^h, T_{\tilde{m}\tilde{n}}(2 - \cos x - \cos y)\tilde{e}^h \rangle = \langle e^h, T_{mn}(2 - \cos x - \cos y)e^h \rangle,$$

the conclusion still holds. $\square$

8.3 Convergence result for full MGM

In the TGM the matrix T^H on the coarse grid is inverted exactly. In the MGM the matrix T^H is usually not solved exactly, but is approximated by using the idea of the TGM recursively on each coarser grid until we get to the coarsest grid. On the coarsest grid, the operator is inverted exactly. In Section 8.2, we have the convergence result of the TGM for the first level. To establish the convergence result of the MGM we need to prove the convergence of the TGM on coarser levels. Recall that on the coarser grid the operator T^H is defined by the Galerkin algorithm (8.1), i.e.,

$$T^H = I_h^H T^h I_H^h.$$

For $T^h \in \mathbb{R}^{m_h n_h \times m_h n_h}$, we note that if

$$m_h = 2c_m + 1, \qquad n_h = 2c_n + 1$$

for some $c_m, c_n \in \mathbb{N}$ then T^H is still a BTTB matrix.

Thus we will only consider the case of BTTB matrices T_{mn} where

$$m = 2^k - 1, \qquad n = 2^l - 1$$

for some $k, l \in \mathbb{N}$. For then, on each level $2 \leq j \leq q$, we have

$$m_j = 2k_j + 1, \qquad n_j = 2l_j + 1$$

for some $k_j, l_j \in \mathbb{N}$. Hence the coarse-grid operators T^j will still be BTTB matrices for $2 \leq j \leq q$. Recall that from the proof of Theorem 8.3, (8.16) implies (8.11). We now prove that if (8.16) holds at a finer level, it holds on the next coarser level when the same interpolation operator is used.

Theorem 8.4 *Let $t_{0,0}^h$ and $t_{0,0}^H$ be the main diagonal entries of T^h and T^H, respectively. Let the interpolation operator $I_H^h = (I_h^H)^T$ be defined as in (8.4) with P given by (8.5) or (8.6). Suppose that*

$$T^h \geq \frac{2t_{0,0}^h}{\beta^h} T_{m_h n_h}(2 \mp \cos x \mp \cos y), \tag{8.17}$$

for some $\beta^h > 0$ independent of mn. Then

$$T^H \geq \frac{2t_{0,0}^H}{\beta^H} T_{m_H n_H}(2 - \cos x - \cos y) \tag{8.18}$$

with

$$\beta^H = \frac{t_{0,0}^H \beta^h}{8t_{0,0}^h}. \tag{8.19}$$

Proof: We first define a matrix $Q_k \in \mathbb{R}^{k \times (k+1)}$ as follows,

$$Q_k = \frac{1}{2} \begin{pmatrix} 1 & 1 & & & \\ & 1 & 1 & & \\ & & \ddots & \ddots & \\ & & & 1 & 1 \end{pmatrix}. \tag{8.20}$$

By noting (8.4), (8.5), (8.6) and (8.20), there exists a permutation matrix B such that

$$B I_H^h = 4 \begin{pmatrix} I_{m_H} \\ \pm Q_{m_H}^T \end{pmatrix} \otimes \begin{pmatrix} I_{n_H} \\ \pm Q_{n_H}^T \end{pmatrix}. \tag{8.21}$$

Moreover, by using the same permutation matrix B, we have

$$BT_{m_h n_h}(2 \mp \cos x \mp \cos y)B$$

$$= \begin{pmatrix} I_{m_H} & \mp Q_{m_H} \\ \mp Q_{m_H}^T & I_{m_H+1} \end{pmatrix} \otimes I_{n_h} + I_{m_h} \otimes \begin{pmatrix} I_{n_H} & \mp Q_{n_H} \\ \mp Q_{n_H}^T & I_{n_H+1} \end{pmatrix}. \tag{8.22}$$

By (8.1) and (8.17) we obtain

$$T^H = I_h^H T^h I_H^h \geq \frac{2 l_{0,0}^h}{\beta^h} I_h^H T_{m_h n_h}(2 \mp \cos x \mp \cos y) I_H^h. \tag{8.23}$$

For $I_h^H T_{m_h n_h}(2 \mp \cos x \mp \cos y) I_H^h$, we have by (8.21) and (8.22),

$$I_h^H T_{m_h n_h}(2 \mp \cos x \mp \cos y) I_H^h$$

$$= 16 \left[(I_{m_H} - Q_{m_H} Q_{m_H}^T) \otimes (I_{n_H} + Q_{n_H} Q_{n_H}^T) \right.$$

$$\left. + (I_{m_H} + Q_{m_H} Q_{m_H}^T) \otimes (I_{n_H} - Q_{n_H} Q_{n_H}^T) \right]$$

$$= 4 T_{m_H n_H}((1 - \cos x)(3 + \cos y))$$

$$+ 4 T_{m_H n_H}((3 + \cos x)(1 - \cos y))$$

$$= 8 T_{m_H n_H}(3 - \cos x - \cos y - \cos x \cos y)$$

$$\geq 8 T_{m_H n_H}(2 - \cos x - \cos y).$$

We then obtain

$$\frac{2t_{0,0}^h}{\beta^h} I_h^H T_{m_h n_h} (2 \mp \cos x \mp \cos y) I_H^h$$

$$\geq \frac{16t_{0,0}^h}{\beta^h} T_{m_H n_H} (2 - \cos x - \cos y).$$

Hence (8.23) implies (8.18) and (8.19).   □

From Theorem 8.4 we conclude that if the first level matrix T_{mn} satisfies (8.2) or (8.3), all coarser level matrices T^j, for $2 \leq j \leq q$, will only satisfy (8.2). Thus we only use the interpolation operator defined as in (8.4) and (8.5) for all coarser levels. Recall by (8.9) that we can choose α^h such that

$$\alpha^h T^h \leq t_{0,0}^h I_{m_h n_h}.$$

Notice that $Q_k Q_k^T \leq I_k$ and therefore we have

$$\alpha^h T^H = \alpha^h I_h^H T^h I_H^h \leq t_{0,0}^h I_h^H I_H^h$$

$$= 16 t_{0,0}^h (I_{m_H} + Q_{m_H} Q_{m_H}^T) \otimes (I_{n_H} + Q_{n_H} Q_{n_H}^T)$$

$$\leq 64 t_{0,0}^h I_{m_H n_H}.$$

Thus, at the coarser level we can choose α^H as

$$\alpha^H = \frac{\alpha^h t_{0,0}^H}{64 t_{0,0}^h}. \tag{8.24}$$

According to Theorem 8.2, (8.24) and (8.19), we see that

$$\|G^H\|_{\bar 1} \leq \sqrt{1 - \frac{\alpha^H}{\beta^H}} = \sqrt{1 - \frac{\alpha^h t_{0,0}^H / (64 t_{0,0}^h)}{t_{0,0}^H \beta^h / (8 t_{0,0}^h)}} = \sqrt{1 - \frac{\alpha^h}{8\beta^h}}.$$

Recursively we obtain the level-dependent convergence rate of the MGM:

$$\|G^q\|_{\bar 1} \leq \sqrt{1 - \frac{\alpha^q}{\beta^q}} = \sqrt{1 - \frac{\alpha^h}{8^{q-1}\beta^h}}.$$

We remark that numerical results in Section 8.4 show that the MGM with V-cycle has already given a level-independent convergence rate.

8.4 Numerical results

Let us first compute the operation cost of the MGM. We only consider the case of $m = 2^k - 1$ and $n = 2^l - 1$ for some $k, l \in \mathbb{N}$. On each level j, we know that $m_j = 2k_j + 1$ and $n_j = 2l_j + 1$ for some $k_j, l_j \in \mathbb{N}$. From the algorithm of the MGM in Section 8.1, we see that if we use the damped Jacobi method (8.8), the pre-smoothing and post-smoothing steps become

$$u^j := (I_{m_j n_j} - \omega(\delta(T^j))^{-1}T^j)u^j + \omega(\delta(T^j))^{-1}b^j.$$

Thus the main cost on each level depends on the matrix-vector multiplication $T^j y$ for some vector y. If we use one pre-smoothing step and one post-smoothing step, then it requires two matrix-vector multiplications — one from the post-smoothing and the other one from the computation of the residual. We do not need the multiplication in the pre-smoothing step if the initial guess u^j is taken to be the zero vector. At the finest level, since $T^1 = T_{mn}$ is a BTTB matrix, $T_{mn}y$ can be computed in two FFTs with $4mn$-length, see Section 2.2.2. At each level j, $T^j y$ can be computed in two FFTs with $4m_j n_j$-length. Thus the total cost per iteration of the MGM is about $16/3$ FFTs with $4mn$-length. Therefore the cost per iteration of the MGM is about $8/3$ times as that required by the conjugate gradient method.

Now, we apply the algorithm of the MGM in Section 8.1 to the ill conditioned BTTB systems

$$T_{mn}[f_i(x, y)]u = b$$

with six different generating functions $f_i(x, y)$ for $i = 1, \cdots, 6$. They are:

(i) $f_1(x, y) = 2 - \cos x - \cos y$;

(ii) $f_2(x, y) = |x| + |y|$;

(iii) $f_3(x, y) = x^2 + y^2$;

(iv) $f_4(x, y) = 20 - 8 \cos x - 8 \cos y - 4 \cos x \cos y$

where $f_4(x, y)$ is the generating function corresponding to the Laplace operator defined on the unit square and discretized by the 9-point stencil as follows (see [32, 74]),

$$\begin{bmatrix} -1 & -4 & -1 \\ -4 & 20 & -4 \\ -1 & -4 & -1 \end{bmatrix} ;$$

(v) $f_5(x, y) = x^2 + y^2 - |xy|;$

(vi) $f_6(x, y) = J(x) + J(y)$

where $J(x)$ is a function with jump:

$$J(x) = \begin{cases} x^2, & \text{if} \quad |x| < \pi/2, \\ 1, & \text{if} \quad |x| \geq \pi/2. \end{cases}$$

We choose a random vector u as a solution such that its components satisfy

$$0 \leq (u)_i \leq 1.$$

The right hand side vector b is obtained accordingly. As smoother, we use the damped Jacobi method with

$$\omega = a_{0,0}/\max f(x, y)$$

for pre-smoother, and

$$\omega = 2a_{0,0}/\max f(x, y)$$

for post-smoother. We use one pre-smoothing and one post-smoothing at each level. The zero vector is the initial guess. The stopping tolerance is

$$\frac{\|u - u^{(j)}\|_\infty}{\|u\|_\infty} < 10^{-4}$$

where $u^{(j)}$ is the approximated solution after the j-th V-cycle. In Tables 8.1 and 8.2 we give the number of iterations for six systems. The sizes of the matrices of the systems are given by $m = n = 2^k - 1$ where k is from 3 to 8. When the size of the coarse grid operator is 9×9, we solve the system exactly.

For a comparison, we also give the number of iterations by the PCG method with no preconditioner I, the preconditioners $S = s_{F,F}^{(2)}(T_{mn})$ and

$C = c_{F,F}^{(2)}(T_{mn})$ proposed in Chapter 2. The '∗' in Tables 8.1 and 8.2 means that the preconditioner is not well-defined. Finally, we remark that the number of iterations of the MGM remains roughly a constant, i.e., our method gives a level-independent convergence rate.

Table 8.1. Number of iterations for three non-negative functions

$f(x,y)$	$f_1(x,y)$				$f_2(x,y)$				$f_3(x,y)$			
k	I	S	C	MGM	I	S	C	MGM	I	S	C	MGM
3	16	∗	8	6	11	7	6	6	20	29	10	11
4	33	∗	11	7	17	8	7	8	45	20	13	12
5	62	∗	16	7	25	9	8	8	90	26	19	12
6	120	∗	23	7	37	10	9	9	180	33	29	12
7	299	∗	36	7	47	12	10	9	352	43	46	12
8	461	∗	56	7	65	13	12	9	715	53	72	12

Table 8.2. Number of iterations for three non-negative functions

$f(x,y)$	$f_4(x,y)$				$f_5(x,y)$				$f_6(x,y)$			
k	I	S	C	MGM	I	S	C	MGM	I	S	C	MGM
3	14	∗	8	5	18	10	11	10	10	10	7	15
4	27	∗	11	5	40	13	15	10	21	10	9	17
5	51	∗	16	6	83	17	23	10	43	14	13	17
6	98	∗	24	6	155	26	35	10	85	19	19	17
7	187	∗	38	6	301	35	56	11	170	33	29	18
8	376	∗	58	6	607	46	92	11	350	63	44	18

Chapter 9

Applications in Second-Order PDEs

In this chapter we first review some results, obtained in [17], which are related to numerical solutions of elliptic boundary value problems. We then consider linear systems arising from implicit time discretizations and finite difference space discretizations of second-order hyperbolic equations. The PCG method with BCCB preconditioners is proposed to solve the linear systems. The results are extended to parabolic equations. Preconditioners based on the fast sine transform are also discussed for discretized systems of second-order partial differential equations in 3-dimensional space. For a literature on iterative solvers for second-order partial differential equations, we refer to [17, 51, 59, 82].

9.1 Applications to elliptic problems

We want to discuss the elliptic equation of the form

$$\frac{\partial}{\partial x_1}\left(a\frac{\partial u}{\partial x_1}\right) + \frac{\partial}{\partial x_2}\left(b\frac{\partial u}{\partial x_2}\right) = f \tag{9.1}$$

defined on the unit square with given Dirichlet boundary condition. Here

$$a = a(x_1, x_2), \quad b = b(x_1, x_2), \quad f = f(x_1, x_2)$$

are given functions with

$$0 < c_{\min} \leq a(x_1, x_2), \ b(x_1, x_2) \leq c_{\max} \tag{9.2}$$

for some constants $c_{\min}$ and $c_{\max}$. Usually the numerical solution of a 2-dimensional elliptic equation on a uniform grid, using the five-point centered difference scheme, involves the solution of block tri-diagonal system of equations $Au = d$. The important properties for the matrix A are its sparsity and bandwidth. If the grid has m interior grid points in each direction, then the block tri-diagonal matrix $A \in \mathbb{R}^{m^2 \times m^2}$ contains only about $5m^2$ non-zero entries. The bandwidth of the A is $2m + 1$. It is desirable to retain the sparsity in solving procedure, and therefore iterative methods are proposed. The conjugate gradient method will be used for solving $Au = d$.

Typically, the convergence rate of the conjugate gradient method depends on the condition number $\kappa(A)$ of the coefficient matrix A. For elliptic problems of second-order, usually $\kappa(A) = O(m^2)$ and hence grows rapidly with m. We then use a preconditioner to alleviate this problem. A successful type of preconditioner is the modified incomplete LU (MILU) factorizations, see [38]. We note that though the conjugate gradient method is highly parallelizable, see [76], both the computation and the application of the MILU preconditioner have limited degree of parallelism because of the inherently sequential way in which the grid points are ordered.

R. Chan and T. Chan in [17] then proposed another class of preconditioners. The preconditioners are based on averaging the coefficients of A to form a circulant approximation to A. They showed that the condition number of the preconditioned system is bounded by $O(m)$. Note that both the computation (based on averaging of the coefficients of the elliptic operator) and the inversion (using FFT's) of these preconditioners are highly parallelizable, see [90].

We now define BCCB preconditioners for elliptic operators on rectangular domains. For (9.1) let the domain be discretized by using a uniform grid with m grid points in each coordinate direction. Let $u_{i,j}$ denote the calculated approximate solution of u at point $(x_{1,i}, x_{2,j})$ and $a_{i,j}, b_{i,j}, f_{i,j}$ denote the values of a, b, f at point $(x_{1,i}, x_{2,j})$ respectively, where

$$\begin{cases} x_{1,i} = ih, & i = 0, \cdots, m+1, \\ x_{2,j} = jh, & j = 0, \cdots, m+1, \end{cases}$$

and h is a step. By using the 5-point centered difference approximation, we then have

$$(a_{i+\frac{1}{2},j} + a_{i-\frac{1}{2},j} + b_{i,j+\frac{1}{2}} + b_{i,j-\frac{1}{2}})u_{i,j} - a_{i+\frac{1}{2},j}u_{i+1,j} - a_{i-\frac{1}{2},j}u_{i-1,j}$$

$$-b_{i,j+\frac{1}{2}}u_{i,j+1} - b_{i,j-\frac{1}{2}}u_{i,j-1} = h^2 f_{i,j}.$$

The values $u_{i,0}$, $u_{i,m+1}$, $u_{0,j}$ and $u_{m+1,j}$, for $i, j = 0, \cdots, m + 1$, are given directly by the boundary condition. This implies that we have to solve for m^2 unknowns.

Let

$$u \equiv (u_{1,1}, u_{2,1}, \cdots, u_{m,1}, u_{1,2}, \cdots, u_{1,m}, \cdots, u_{m,m})^T$$

and

$$d \equiv h^2(f_{1,1}, f_{2,1}, \cdots, f_{m,1}, f_{1,2}, \cdots, f_{1,m}, \cdots, f_{m,m})^T,$$

then we finally obtain the following linear system

$$Au = d. \tag{9.3}$$

Here $A \in \mathbb{R}^{m^2 \times m^2}$ is a block tri-diagonal matrix where diagonal blocks are tri-diagonal matrices and off-diagonal blocks are diagonal matrices.

We consider the modified optimal BCCB preconditioner which preserves the block structure of A. The preconditioner C is defined as follows:

$$C = I \otimes C^a + C^b \otimes I. \tag{9.4}$$

Here $I \in \mathbb{R}^{m \times m}$ is the identity matrix and C^a, C^b are circulant matrices in $\mathbb{R}^{m \times m}$ with their first columns defined by:

$$\begin{cases} c_0^a = 2\bar{a} + \dfrac{1}{m^2}, \qquad c_1^a = c_{m-1}^a = -\bar{a}, \\[2mm] c_i^a = 0, \quad i = 2, \cdots, m-2; \\[2mm] c_0^b = 2\bar{b} + \dfrac{1}{m^2}, \qquad c_1^b = c_{m-1}^b = -\bar{b}, \\[2mm] c_i^b = 0, \quad i = 2, \cdots, m-2, \end{cases}$$

where

$$\bar{a} = \frac{1}{m^2} \sum_{j=1}^{m} \sum_{i=1}^{m-1} a_{i+\frac{1}{2},j} \quad \text{and} \quad \bar{b} = \frac{1}{m^2} \sum_{i=1}^{m} \sum_{j=1}^{m-1} b_{i,j+\frac{1}{2}}.$$

The shift $1/m^2$ can guarantee the reduction of the condition number for the preconditioned system. The proof of the following theorem can be found in [17].

Theorem 9.1 *Let A be the matrix defined as in (9.3) of the equation (9.1) under the condition (9.2). Let C be the BCCB preconditioner of A defined as in (9.4). Then we have*

$$O(1) \le \lambda(C^{-1}A) \le O(m).$$

As a consequence we have

$$\kappa(C^{-1}A) \le O(m).$$

Numerical results in [17] show that the preconditioned systems often have clustering of eigenvalues, which is favorable to the convergence rate.

9.2 Applications to hyperbolic problems

In this section we extend the idea of using BCCB preconditioners in Section 9.1 to hyperbolic equations. We discuss the second-order hyperbolic equation of the following form

$$\frac{\partial^2 z}{\partial t^2} = \frac{\partial}{\partial x_1}\left(a\frac{\partial z}{\partial x_1}\right) + \frac{\partial}{\partial x_2}\left(b\frac{\partial z}{\partial x_2}\right) + g$$

defined on the unit square with given initial and Dirichlet boundary conditions. A number of common methods for solving such kind of problem are explicit finite difference schemes, see [66]. For explicit methods, however, the maximal time step $k_{\max}$ is limited by the CFL criterion, which in some situations may be unrealistically strict. An example is when the time-dependence of the problem is much weaker than the space-dependence, and hence a large time step could be used.

An alternative way is to use implicit schemes. Usually, the numerical solution of 2-dimensional second-order hyperbolic equation on a uniform grid, using an implicit time-marching scheme, involves the solution of block tri-diagonal system of equations in each time step. As in the elliptic case, the important properties for the matrices of such systems are their sparsity and bandwidth. We thus consider the use of the PCG method with BCCB preconditioners. The main result in this section is that for second-order hyperbolic equations defined on the unit square with given initial and Dirichlet boundary conditions, the condition number of the pre-conditioned system is of $O(\alpha)$ or $O(m)$, where α is the grid ratio between the time step and space step, and m is the number of interior gridpoints in each direction.

9.2.1   Circulant approximation to system

We consider the following second-order hyperbolic equation

$$\frac{\partial^2 z}{\partial t^2} = \frac{\partial}{\partial x_1}\left(a\frac{\partial z}{\partial x_1}\right) + \frac{\partial}{\partial x_2}\left(b\frac{\partial z}{\partial x_2}\right) + g \tag{9.5}$$

where $0 < x_1 < 1$, $0 < x_2 < 1$, $t > 0$ and

$$a = a(x_1, x_2), \quad b = b(x_1, x_2), \quad g = g(t, x_1, x_2)$$

are given functions with

$$0 < c_{\min} \leq a(x_1, x_2),\ b(x_1, x_2) \leq c_{\max} \tag{9.6}$$

for some constants $c_{\min}$ and $c_{\max}$. The initial conditions are given by

$$z(0, x_1, x_2) = f_0(x_1, x_2), \quad z_t(0, x_1, x_2) = f_1(x_1, x_2),$$

and the boundary conditions are given by

$$z(t, 0, x_2) = z_0(t, x_2), \quad z(t, 1, x_2) = z_1(t, x_2),$$

$$z(t, x_1, 0) = z_2(t, x_1), \quad z(t, x_1, 1) = z_3(t, x_1). \tag{9.7}$$

In this way we obtain a mixed initial and boundary value problem.

Let

$$u = z_t, \quad w = az_{x_1}, \quad v = bz_{x_2},$$

then we have the following first-order system

$$
\begin{cases}
\dfrac{\partial u}{\partial t} = \dfrac{\partial w}{\partial x_1} + \dfrac{\partial v}{\partial x_2} + g, \\[2mm]
\dfrac{\partial w}{\partial t} = a\dfrac{\partial u}{\partial x_1}, \\[2mm]
\dfrac{\partial v}{\partial t} = b\dfrac{\partial u}{\partial x_2}.
\end{cases}
\tag{9.8}
$$

The grid is uniform in the computational domain with $(m + 2) \times (m + 2)$ gridpoints, where $m \geq 2$. Let $u_{i,j}, v_{i,j}, w_{i,j}$ denote the calculated approximate solutions of u, v, w at point $(x_{1,i}, x_{2,j})$ and $a_{i,j}, b_{i,j}, g_{i,j}$ denote the values of a, b, g at point $(x_{1,i}, x_{2,j})$ respectively, where

$$
\begin{cases}
x_{1,i} = ih, & i = 0, \cdots, m + 1, \\
x_{2,j} = jh, & j = 0, \cdots, m + 1,
\end{cases}
$$

and h is a space step. By using the trapezoidal rule with a time step k to do the time discretization of (9.8), and then followed by using the centered difference scheme to approximate spatial derivatives, we then have

$$
\begin{cases}
\dfrac{u_{i,j}^{n+1} - u_{i,j}^n}{k} - \dfrac{w_{i+\frac{1}{2},j}^{n+1} - w_{i-\frac{1}{2},j}^{n+1} + w_{i+\frac{1}{2},j}^n - w_{i-\frac{1}{2},j}^n}{2h}, \\[3mm]
-\dfrac{v_{i,j+\frac{1}{2}}^{n+1} - v_{i,j-\frac{1}{2}}^{n+1} + v_{i,j+\frac{1}{2}}^n - v_{i,j-\frac{1}{2}}^n}{2h} = \dfrac{1}{2}(g_{i,j}^{n+1} + g_{i,j}^n), \\[3mm]
\dfrac{w_{i-\frac{1}{2},j}^{n+1} - w_{i-\frac{1}{2},j}^n}{k} - a_{i-\frac{1}{2},j}\dfrac{u_{i,j}^{n+1} - u_{i-1,j}^{n+1} + u_{i,j}^n - u_{i-1,j}^n}{2h} = 0, \\[3mm]
\dfrac{v_{i,j-\frac{1}{2}}^{n+1} - v_{i,j-\frac{1}{2}}^n}{k} - b_{i,j-\frac{1}{2}}\dfrac{u_{i,j}^{n+1} - u_{i,j-1}^{n+1} + u_{i,j}^n - u_{i,j-1}^n}{2h} = 0.
\end{cases}
\tag{9.9}
$$

Let $\alpha = k/h$ and substitute the last two equations in (9.9) into the first one, we have

$$
\left(\frac{4}{\alpha^2} + a_{i+\frac{1}{2},j} + a_{i-\frac{1}{2},j} + b_{i,j+\frac{1}{2}} + b_{i,j-\frac{1}{2}}\right)u_{i,j}^{n+1}
$$

$$
-a_{i+\frac{1}{2},j}u_{i+1,j}^{n+1} - a_{i-\frac{1}{2},j}u_{i-1,j}^{n+1}
$$

$$
-b_{i,j+\frac{1}{2}}u_{i,j+1}^{n+1} - b_{i,j-\frac{1}{2}}u_{i,j-1}^{n+1} = \frac{d_{i,j}^{n+1}}{\alpha^2}
$$

where $d_{i,j}^{n+1}$ are known quantities. The values $u_{i,0}^{n+1}$, $u_{i,m+1}^{n+1}$, $u_{0,j}^{n+1}$ and $u_{m+1,j}^{n+1}$, for $i, j = 0, \cdots, m + 1$, are given directly by the boundary conditions (9.7). This implies that we have to solve for m^2 unknowns in each time step.

Let

$$u^{n+1} \equiv (u_{1,1}^{n+1}, u_{2,1}^{n+1}, \cdots, u_{m,1}^{n+1}, u_{1,2}^{n+1}, \cdots, u_{1,m}^{n+1}, \cdots, u_{m,m}^{n+1})^T$$

and

$$d^{n+1} \equiv \frac{1}{\alpha^2}(d_{1,1}^{n+1}, d_{2,1}^{n+1}, \cdots, d_{m,1}^{n+1}, d_{1,2}^{n+1}, \cdots, d_{1,m}^{n+1}, \cdots, d_{m,m}^{n+1})^T,$$

then we finally obtain the following linear system

$$Au^{n+1} = d^{n+1}. \tag{9.10}$$

Here $A \in \mathbb{R}^{m^2 \times m^2}$ is a block tri-diagonal matrix where diagonal blocks are tri-diagonal matrices and off-diagonal blocks are diagonal matrices. Once we get u^{n+1}, we could obtain $z_{i,j}^{n+1}$, the approximation of $z(t_{n+1}, x_{1,i}, x_{2,j})$, by the following difference scheme

$$u_{i,j}^{n+1} = \frac{z_{i,j}^{n+1} - z_{i,j}^n}{k},$$

i.e.,

$$z_{i,j}^{n+1} = ku_{i,j}^{n+1} + z_{i,j}^n.$$

Hence we only need to discuss the solution of (9.10).

We introduce the following BCCB preconditioner which preserves the block structure of A. The preconditioner C is defined as follows:

$$C = I \otimes C^a + C^b \otimes I. \tag{9.11}$$

Here $I \in \mathbb{R}^{m \times m}$ is the identity matrix and C^a, C^b are circulant matrices in $\mathbb{R}^{m \times m}$ with their first columns defined by:

$$\begin{cases} c_0^a = 2\bar{a} + \dfrac{2\beta}{\alpha^2} + \dfrac{1}{m^2}\left(1 + \dfrac{1}{\alpha^2}\right), \\[2mm] c_1^a = c_{m-1}^a = -\bar{a}, \\[2mm] c_i^a = 0, \quad i = 2, \cdots, m-2; \\[2mm] c_0^b = 2\bar{b} + \dfrac{2\beta}{\alpha^2} + \dfrac{1}{m^2}\left(1 + \dfrac{1}{\alpha^2}\right), \\[2mm] c_1^b = c_{m-1}^b = -\bar{b}, \\[2mm] c_i^b = 0, \quad i = 2, \cdots, m-2, \end{cases}$$

where

$$\bar{a} = \frac{1}{m^2} \sum_{j=1}^{m} \sum_{i=1}^{m-1} a_{i+\frac{1}{2},j}, \qquad \bar{b} = \frac{1}{m^2} \sum_{i=1}^{m} \sum_{j=1}^{m-1} b_{i,j+\frac{1}{2}},$$

$$\beta = (m-1)/m \quad \text{and} \quad \alpha = k/h.$$

The shift

$$\frac{1}{m^2}\left(1 + \frac{1}{\alpha^2}\right)$$

can guarantee the reduction of the condition number for the preconditioned system.

The question that we are facing now is that how good this preconditioner is in the sense of minimizing the condition number $\kappa(C^{-1}A)$. We will show that

1. For any α, when m is sufficiently large ($m \gg \alpha$),

$$\kappa(C^{-1}A) \le O(\alpha),$$

   whilest for the original matrix, $\kappa(A) \le O(\alpha^2)$.

2. For any m, when α is sufficiently large ($\alpha \gg m$),

$$\kappa(C^{-1}A) \le O(m),$$

   whilest for the original matrix, $\kappa(A) \le O(m^2)$.

We first prove the claims above for a model problem in case of

$$a(x_1, x_2) \equiv b(x_1, x_2) \equiv 1$$

in Section 9.2.2 and then extend the results to the general variable-coefficient case in Section 9.2.3.

9.2.2   Analysis for model problem

In the constant-coefficient case of $a(x_1, x_2) \equiv b(x_1, x_2) \equiv 1$, $A \in \mathbb{R}^{m^2 \times m^2}$ is of the following form

$$A = A_0 \otimes I + I \otimes A_0 \tag{9.12}$$

where $A_0 \in \mathbb{R}^{m \times m}$ is given by

$$A_0 = \begin{pmatrix} 2 + 2/\alpha^2 & -1 & & 0 \\ -1 & \ddots & \ddots & \\ & \ddots & \ddots & -1 \\ 0 & & -1 & 2 + 2/\alpha^2 \end{pmatrix}.$$

In this case,

$$\bar{a} = \bar{b} = \beta = \frac{m-1}{m}.$$

In particular, the BCCB preconditioner $C \in \mathbb{R}^{m^2 \times m^2}$ is given by

$$C = C_0 \otimes I + I \otimes C_0 \tag{9.13}$$

where $C_0 \in \mathbb{R}^{m \times m}$ is given by

$$C_0 = \beta \begin{pmatrix} 2 + 2/\alpha^2 & -1 & & & -1 \\ -1 & \ddots & \ddots & 0 & \\ & \ddots & \ddots & \ddots & \\ & 0 & \ddots & \ddots & -1 \\ -1 & & & -1 & 2 + 2/\alpha^2 \end{pmatrix} + \frac{1}{m^2}(1 + \frac{1}{\alpha^2})I.$$

Hence C_0 is a positive definite circulant matrix. For the eigenvalues of A_0 and C_0, we have the following lemma.

Lemma 9.1 *The eigenvalues of A_0 and C_0 are given as follows:*

$$\lambda_j(A_0) = \frac{2}{\alpha^2} + 4\sin^2\frac{\pi(j+1)}{2m+2}, \tag{9.14}$$

$$\lambda_j(C_0) = \frac{2\beta}{\alpha^2} + \frac{1}{m^2}(1 + \frac{1}{\alpha^2}) + 4\beta\sin^2\frac{\pi j}{m}, \tag{9.15}$$

for $j = 0, \cdots, m-1$.

Proof: For (9.14) one can refer to [66]. For (9.15) since C_0 is a circulant matrix, we have $C_0 = F_m^* \Lambda_m F_m$, see (1.6). By using this spectral decomposition, one can easily obtain (9.15). □

By (9.12) and (9.14), we know that the eigenvalues of A are given by

$$\lambda_{i,j}(A) = \frac{4}{\alpha^2} + 4\sin^2\frac{\pi(i+1)}{2m+2} + 4\sin^2\frac{\pi(j+1)}{2m+2}, \tag{9.16}$$

for $0 \leq i, j \leq m-1$. From (9.16), we know that when α is sufficiently large ($\alpha \gg m$), then the smallest eigenvalue of A decreases to zero like $O(1/m^2)$. Since $\lambda_{i,j}(A) \leq 9$, for $\alpha \geq 4$ and $0 \leq i, j \leq m-1$, as a consequence, we have

$$\kappa(A) \leq O(m^2) \ .$$

If m is sufficiently large ($m \gg \alpha$), then the smallest eigenvalue of A decreases to zero like $O(1/\alpha^2)$. As a consequence we have

$$\kappa(A) \leq O(\alpha^2) \ .$$

For the condition number of $C_0^{-1}A_0$, we have the following two lemmas.

Lemma 9.2 *Let $\lambda(C_0^{-1}A_0)$ denote any eigenvalue of $C_0^{-1}A_0$. For any α, when m is sufficiently large ($m \gg \alpha$), we have*

$$\frac{1}{2} \leq \lambda(C_0^{-1}A_0) \leq O(\alpha).$$

As a consequence we have

$$\kappa(C_0^{-1}A_0) \leq O(\alpha), \qquad \text{when } m \gg \alpha.$$

Proof: Let $e_j \in \mathbb{R}^m$ be the j-th unit vector. Since

$$C_0 = \beta(A_0 - e_1 e_m^T - e_m e_1^T) + \frac{1}{m^2}(1 + \frac{1}{\alpha^2})I,$$

we have, for all vectors $x \in \mathbb{R}^m$,

$$x^T C_0 x = \beta x^T A_0 x + \beta x^T (e_1 e_1^T + e_m e_m^T)x$$

$$-\beta x^T (e_1 + e_m)(e_1 + e_m)^T x + \frac{1}{m^2}(1 + \frac{1}{\alpha^2})x^T x$$

$$= 2\beta x^T A_0 x - \beta x^T [A_0 - (e_1 e_1^T + e_m e_m^T) - \frac{1}{\beta m^2}(1 + \frac{1}{\alpha^2})I]x$$

$$-\beta x^T (e_1 + e_m)(e_1 + e_m)^T x.$$

We note that the matrix $(e_1 + e_m)(e_1 + e_m)^T$ is positive semi-definite and the matrix

$$A_0 - (e_1 e_1^T + e_m e_m^T) - \frac{1}{\beta m^2}\left(1 + \frac{1}{\alpha^2}\right)I$$

is also positive semi-definite when $m \geq \sqrt{\alpha^2 + 1}$. We then have for any α, when m is sufficiently large,

$$x^T C_0 x \leq 2\beta x^T A_0 x.$$

Thus

$$\frac{1}{2} \leq \frac{1}{2\beta} \leq \min_{\|x\| \neq 0} \frac{x^T A_0 x}{x^T C_0 x} \leq \lambda(C_0^{-1} A_0).$$

On the other hand, we note that for all vectors $x \in \mathbb{R}^m$,

$$\beta x^T A_0 x = x^T C_0 x + \frac{\beta}{2}x^T (e_1 + e_m)(e_1 + e_m)^T x$$

$$-\frac{\beta}{2}x^T (e_1 - e_m)(e_1 - e_m)^T x - \frac{1}{m^2}(1 + \frac{1}{\alpha^2})x^T x,$$

where the last two terms on the right hand side is always non-positive. Thus

$$\beta x^T A_0 x \leq x^T C_0 x + \frac{\beta}{2}x^T e e^T x$$

where $e = e_1 + e_m$, i.e.,

$$\frac{x^T A_0 x}{x^T C_0 x} \leq \frac{1}{\beta} + \frac{1}{2}\frac{x^T e e^T x}{x^T C_0 x}. \qquad (9.17)$$

We note that for all non-zero vectors $x \in \mathbb{R}^m$,

$$\frac{x^T e e^T x}{x^T C_0 x} \leq \|C_0^{-1/2} e e^T C_0^{-1/2}\|_2 = e^T C_0^{-1} e. \tag{9.18}$$

Since $C_0 = F_m^* \Lambda_m F_m$, by (9.15), the entries of Λ_m are given by

$$[\Lambda_m]_{j,j} = \lambda_j(C_0) = \frac{2\beta}{\alpha^2} + \frac{1}{m^2}(1 + \frac{1}{\alpha^2}) + 4\beta \sin^2 \theta_j$$

where $\theta_j = \pi j/m$ for $0 \leq j \leq m - 1$. Hence

$$e^T C_0^{-1} e = e^T F_m^* \Lambda_m^{-1} F_m e = \frac{4}{m} \sum_{j=0}^{m-1} \frac{\cos^2 \theta_j}{\frac{2\beta}{\alpha^2} + \frac{1}{m^2}(1 + \frac{1}{\alpha^2}) + 4\beta \sin^2 \theta_j}$$

$$= \frac{4m\alpha^2}{2m^2 - 2m + \alpha^2 + 1} + \frac{8}{m} \sum_{j=1}^{m/2-1} \frac{\cos^2 \theta_j}{\frac{2\beta}{\alpha^2} + \frac{1}{m^2}(1 + \frac{1}{\alpha^2}) + 4\beta \sin^2 \theta_j}$$

$$\leq \frac{4m\alpha^2}{2m^2 - 2m + \alpha^2 + 1} + \frac{4}{\pi} \int_{\frac{\pi}{m}}^{\frac{\pi}{2}} \frac{\cos^2 \theta \, d\theta}{\frac{1}{2\alpha^2} + \sin^2 \theta}. \tag{9.19}$$

For any α, when m is sufficiently large ($m \gg \alpha$), we have by (9.19),

$$e^T C_0^{-1} e \leq 1 + \frac{4}{\pi} \int_0^{\frac{\pi}{2}} \frac{\cos^2 \theta \, d\theta}{\frac{1}{2\alpha^2} + \sin^2 \theta}$$

$$\leq 1 + \frac{4 \cos^2 \hat{\theta}}{\pi} \int_0^{\frac{\pi}{2}} \frac{d\theta}{\frac{1}{2\alpha^2} + \sin^2 \theta}$$

$$= 1 + \frac{4 \cos^2 \hat{\theta}}{\pi} \frac{\pi}{2\sqrt{\frac{1}{2\alpha^2}(\frac{1}{2\alpha^2} + 1)}}, \tag{9.20}$$

where $0 < \hat{\theta} < \frac{\pi}{2}$. By (9.18) and (9.20), we then have

$$\frac{x^T e e^T x}{x^T C_0 x} \leq e^T C_0^{-1} e \leq 1 + c\alpha = O(\alpha) \tag{9.21}$$

where c is a constant. By (9.17) and (9.21) we have

$$\lambda(C_0^{-1} A_0) \leq \max_{\|x\| \neq 0} \frac{x^T A_0 x}{x^T C_0 x} \leq O(\alpha). \qquad \square$$

Lemma 9.3 *For any* m, *when* α *is sufficiently large* $(\alpha \gg m)$, *we have*

$$O(1) \leq \lambda(C_0^{-1}A_0) \leq O(m).$$

As a consequence, we have

$$\kappa(C_0^{-1}A_0) \leq O(m), \qquad when \; \alpha \gg m.$$

Proof: We note that

$$x^T C_0 x = 2\beta x^T A_0 x - \beta x^T [A_0 - (e_1 e_1^T + e_m e_m^T)]x$$

$$-\beta x^T (e_1 + e_m)(e_1 + e_m)^T x + \frac{1}{m^2}\left(1 + \frac{1}{\alpha^2}\right)x^T x.$$

Since the matrices $(e_1 + e_m)(e_1 + e_m)^T$ and $A_0 - (e_1 e_1^T + e_m e_m^T)$ are positive semi-definite, we have

$$x^T C_0 x \leq 2\beta x^T A_0 x + \frac{1}{m^2}\left(1 + \frac{1}{\alpha^2}\right)x^T x. \qquad (9.22)$$

When α is sufficiently large $(\alpha \gg m)$ we know that

$$x^T x \leq O(m^2)x^T A_0 x.$$

Using this result, we see from (9.22) that

$$(2\beta + O(1))^{-1} \leq \lambda(C_0^{-1}A_0),$$

i.e.,

$$O(1) \leq \lambda(C_0^{-1}A_0).$$

On the other hand, for any m, when α is sufficiently large $(\alpha \gg m)$, we have by (9.18) and (9.19)

$$\frac{x^T e e^T x}{x^T C_0 x} \leq e^T C_0^{-1} e \leq O(m) + \frac{4}{\pi}\int_{\frac{\pi}{m}}^{\frac{\pi}{2}} \frac{\cos^2\theta d\theta}{\sin^2\theta}$$

$$\leq O(m) + \frac{4}{\pi}\int_{\frac{\pi}{m}}^{\frac{\pi}{2}} \frac{d\theta}{\sin^2\theta} = O(m) + \frac{4}{\pi}\cot(\frac{\pi}{m})$$

$$= O(m). \qquad (9.23)$$

By (9.17) and (9.23), we thus have

$$\lambda(C_0^{-1}A_0) \le O(m). \quad \square$$

By using Lemmas 9.2 and 9.3 we then have

Theorem 9.2 *For the preconditioned system of the model problem, we have*

(i) *For any α, when m is sufficiently large $(m \gg \alpha)$, we have*

$$\frac{1}{2} \le \lambda(C^{-1}A) \le O(\alpha).$$

(ii) *For any m, when α is sufficiently large $(\alpha \gg m)$, we have*

$$O(1) \le \lambda(C^{-1}A) \le O(m).$$

As a consequence we have

$$\kappa(C^{-1}A) \le O(\alpha), \qquad when\ m \gg \alpha;$$

and

$$\kappa(C^{-1}A) \le O(m), \qquad when\ \alpha \gg m.$$

Proof: For (i), we note that for any vector $x \in \mathbb{R}^m$, when $m \ge \sqrt{\alpha^2 + 1}$, by Lemma 9.2,

$$\frac{1}{2}x^T C_0 x \le x^T A_0 x \le O(\alpha)x^T C_0 x.$$

Hence, for any vector $x \in \mathbb{R}^{m^2}$ one can easily prove that

$$\frac{1}{2}x^T(C_0 \otimes I)x \le x^T(A_0 \otimes I)x \le O(\alpha)x^T(C_0 \otimes I)x$$

and

$$\frac{1}{2}x^T(I \otimes C_0)x \le x^T(I \otimes A_0)x \le O(\alpha)x^T(I \otimes C_0)x.$$

Combining these two inequalities together, we have (i). Similarly, we can prove (ii) by using Lemma 9.3. $\quad \square$

For the PCG method it is important that the spectrum of $C^{-1}A$ is clustered in a small interval (a, b) which keeps a clear gap between a and 0, see Theorem 1.9. For $m = 4, 8, 16$, Tables 9.1–9.3 in Section 9.4 show the distributions of the eigenvalues for increasing α. In these tables the eigenvalues are ordered as

$$\lambda_1 \le \lambda_2 \le \cdots \le \lambda_{m^2-1} \le \lambda_{m^2}.$$

We observe that the eigenvalues of $C^{-1}A$ are all located in a relatively small interval (c, d) except one outlying eigenvalue which increases like $O(m)$ for α large enough, just as Theorem 9.2 predicts. Here d is increased slowly with α and m increasing. Since the spectrum of $C^{-1}A$ is clustered, which is favorable to the PCG method, we can expect a fast convergence rate.

9.2.3   Analysis for variable-coefficient problem

We extend the results in Section 9.2.2 to the variable-coefficient case. We consider the second-order hyperbolic equation given by (9.5). Let $\widetilde{A} \in \mathbb{R}^{m^2 \times m^2}$ be given by (9.10). Define

$$A_{\max} = c_{\max} \cdot A \quad \text{and} \quad A_{\min} = c_{\min} \cdot A,$$

where $c_{\max}, c_{\min}$ are given in (9.6) and A is given by (9.12). Without loss of generality, we assume

$$c_{\min} \le 1 \quad \text{and} \quad c_{\max} \ge 1.$$

Let $\widetilde{C}$, $C_{\max}$ and $C_{\min}$ be the BCCB approximations of $\widetilde{A}$, $A_{\max}$ and $A_{\min}$ respectively. Clearly,

$$C_{\max} = c_{\max} \cdot C \quad \text{and} \quad C_{\min} = c_{\min} \cdot C$$

where C is given by (9.13). We then have the following lemma. The proof of the lemma is straightforward.

Lemma 9.4 *All the matrices $A_{\max} - \widetilde{A}$, $\widetilde{A} - A_{\min}$, $C_{\max} - \widetilde{C}$ and $\widetilde{C} - C_{\min}$ are positive semi-definite.*

By Lemma 9.4, for all non-zero vectors $x \in \mathbb{R}^{m^2}$, we have

$$0 < x^T A_{\min} x \leq x^T \tilde{A} x \leq x^T A_{\max} x \tag{9.24}$$

and

$$0 < x^T C_{\min} x \leq x^T \tilde{C} x \leq x^T C_{\max} x. \tag{9.25}$$

Combining (9.24) with (9.25), we get

$$0 < \frac{c_{\min}}{c_{\max}} \frac{x^T A x}{x^T C x} = \frac{x^T A_{\min} x}{x^T C_{\max} x} \leq \frac{x^T \tilde{A} x}{x^T \tilde{C} x} \leq \frac{x^T A_{\max} x}{x^T C_{\min} x} = \frac{c_{\max}}{c_{\min}} \frac{x^T A x}{x^T C x}.$$

Recalling the results from Theorem 9.2, we then have the main results in this section.

Theorem 9.3 *Let $\tilde{A}$ be the discretization matrix of (9.5) defined by (9.10) under the condition (9.6) and $\tilde{C}$ be the BCCB preconditioner defined as in (9.11). We have*

(i) *For any α, when m is sufficiently large ($m \gg \alpha$) we have*

$$O(1) \leq \lambda(\tilde{C}^{-1}\tilde{A}) \leq O(\alpha).$$

(ii) *For any m, when α is sufficiently large ($\alpha \gg m$) we have*

$$O(1) \leq \lambda(\tilde{C}^{-1}\tilde{A}) \leq O(m).$$

As a consequence we have

$$\kappa(\tilde{C}^{-1}\tilde{A}) \leq O(\alpha), \qquad \text{when } m \gg \alpha;$$

and

$$\kappa(\tilde{C}^{-1}\tilde{A}) \leq O(m), \qquad \text{when } \alpha \gg m.$$

9.3 Extension to parabolic equations

In this section, we extend our results to parabolic equations. We consider the following parabolic equation

$$\frac{\partial z}{\partial t} = \frac{\partial}{\partial x_1}\left(a\frac{\partial z}{\partial x_1}\right) + \frac{\partial}{\partial x_2}\left(b\frac{\partial z}{\partial x_2}\right) + g \tag{9.26}$$

where $0 < x_1 < 1$, $0 < x_2 < 1$, $t > 0$ and

$$a = a(x_1, x_2), \quad b = b(x_1, x_2), \quad g = g(t, x_1, x_2)$$

are given functions with

$$0 < c_{\min} \leq a(x_1, x_2), \ b(x_1, x_2) \leq c_{\max} \qquad (9.27)$$

for some constants $c_{\min}$ and $c_{\max}$. The initial condition is given by

$$z(0, x_1, x_2) = g_0(x_1, x_2),$$

and the boundary conditions are given by

$$z(t, 0, x_2) = z_0(t, x_2), \quad z(t, 1, x_2) = z_1(t, x_2),$$

$$z(t, x_1, 0) = z_2(t, x_1), \quad z(t, x_1, 1) = z_3(t, x_1).$$

By using the uniform grid and notations introduced in Section 9.2, for any function $f(x_1, x_2)$ defined on the unit square we define the following forward and backward differences as

$$\triangle_i f_{i,j} = f_{i+1,j} - f_{i,j}, \quad \nabla_i f_{i,j} = f_{i,j} - f_{i-1,j},$$

$$\triangle^j f_{i,j} = f_{i,j+1} - f_{i,j}, \quad \nabla^j f_{i,j} = f_{i,j} - f_{i,j-1}.$$

Then by applying Crank–Nicholson scheme to (9.26), see [63], we have

$$\frac{z_{i,j}^{n+1} - z_{i,j}^n}{k} - \frac{1}{2h^2}[\triangle_i(a_{i-\frac{1}{2},j} \nabla_i z_{i,j}^{n+1}) + \triangle^j(b_{i,j-\frac{1}{2}} \nabla^j z_{i,j}^{n+1})]$$

$$\qquad (9.28)$$

$$- \frac{1}{2h^2}[\triangle_i(a_{i-\frac{1}{2},j} \nabla_i z_{i,j}^n) + \triangle^j(b_{i,j-\frac{1}{2}} \nabla^j z_{i,j}^n)] = g_{i,j}^{n+\frac{1}{2}}.$$

Thus, we have from (9.28)

$$\left(\frac{2}{\alpha^2} + a_{i+\frac{1}{2},j} + a_{i-\frac{1}{2},j} + b_{i,j+\frac{1}{2}} + b_{i,j-\frac{1}{2}} \right) z_{i,j}^{n+1}$$

$$- a_{i+\frac{1}{2},j} z_{i+1,j}^{n+1} - a_{i-\frac{1}{2},j} z_{i-1,j}^{n+1}$$

$$- b_{i,j+\frac{1}{2}} z_{i,j+1}^{n+1} - b_{i,j-\frac{1}{2}} z_{i,j-1}^{n+1} = \frac{d_{i,j}^{n+1}}{\alpha^2}$$

where $\alpha^2 = k/h^2$ and $d_{i,j}^{n+1}$ are known quantities. Finally, we obtain the following linear system

$$Az^{n+1} = d^{n+1} \tag{9.29}$$

with a block tri-diagonal matrix $A \in \mathbb{R}^{m^2 \times m^2}$. Let

$$\bar{a} = \frac{1}{m^2} \sum_{j=1}^{m} \sum_{i=1}^{m-1} a_{i+\frac{1}{2},j} \quad \text{and} \quad \bar{b} = \frac{1}{m^2} \sum_{i=1}^{m} \sum_{j=1}^{m-1} b_{i,j+\frac{1}{2}}.$$

We define the BCCB preconditioner $C \in \mathbb{R}^{m^2 \times m^2}$ as follows:

$$C = I \otimes C^a + C^b \otimes I. \tag{9.30}$$

Here C^a, C^b are circulant matrices in $\mathbb{R}^{m \times m}$ with their first columns defined by:

$$\begin{cases} c_0^a = 2\bar{a} + \dfrac{\beta}{\alpha^2} + \dfrac{1}{m^2}\left(1 + \dfrac{1}{\alpha^2}\right), \\[2mm] c_1^a = c_{m-1}^a = -\bar{a}, \\[2mm] c_i^a = 0, \quad i = 2, \cdots, m-2; \\[2mm] c_0^b = 2\bar{b} + \dfrac{\beta}{\alpha^2} + \dfrac{1}{m^2}\left(1 + \dfrac{1}{\alpha^2}\right), \\[2mm] c_1^b = c_{m-1}^b = -\bar{b}, \\[2mm] c_i^b = 0, \quad i = 2, \cdots, m-2, \end{cases}$$

where $\beta = (m-1)/m$ and $\alpha^2 = k/h^2$. By using the same trick as in Section 9.2, we then have

Theorem 9.4 *Let A be the discretization matrix of (9.26) defined by (9.29) under the condition (9.27) and C be the BCCB preconditioner defined as in (9.30). We have:*

(i) *For any α, when m is sufficiently large ($m \gg \alpha$) we have*

$$O(1) \leq \lambda(C^{-1}A) \leq O(\alpha).$$

(ii) *For any m, when α is sufficiently large ($\alpha \gg m$) we have*

$$O(1) \le \lambda(C^{-1}A) \le O(m).$$

As a consequence we have

$$\kappa(C^{-1}A) \le O(\alpha), \qquad when\ m \gg \alpha;$$

and

$$\kappa(C^{-1}A) \le O(m), \qquad when\ \alpha \gg m.$$

We remark here that the condition number of the original matrix A is of $O(\alpha^2)$ or $O(m^2)$ respectively.

9.4   Numerical results

In this section, we compare the performance of our method to the MILU preconditioner, see [38]. In all tests, we mainly compare the number of iterations. The equations we used are:

$$\frac{\partial^2 z}{\partial t^2} = \frac{\partial}{\partial x}\left[(1 + \epsilon e^{xy})\frac{\partial z}{\partial x}\right] + \frac{\partial}{\partial y}\left[\left(1 + \frac{\epsilon}{2}\cos(\pi x + \pi y)\right)\frac{\partial z}{\partial y}\right] + g(t, x, y)$$
(9.31)

and

$$\frac{\partial z}{\partial t} = \frac{\partial}{\partial x}\left[\left(1 + \frac{\epsilon}{2}\sin(2\pi xy)\right)\frac{\partial z}{\partial x}\right] + \frac{\partial}{\partial y}\left[\left(1 + \epsilon(x^2 + y^2)\right)\frac{\partial z}{\partial y}\right] + g(t, x, y)$$
(9.32)

defined on the unit square. The ϵ here is a parameter. When $\epsilon = 0$, (9.31) is the model problem discussed in Section 9.2.2. We discretize the equations by using the schemes we introduced in Sections 9.2 and 9.3 accordingly. Both the right hand side and the initial guess are chosen to be random vectors and are the same for different methods. The iterations are stopped when

$$\frac{\|r^{(k)}\|_2}{\|r^{(0)}\|_2} < 10^{-7}.$$

Here $r^{(k)}$ is the residual vector at the k-th iteration. The BCCB preconditioner we used is defined by (9.11) or (9.30) accordingly.

Since the BCCB preconditioners are based on averaging of the coefficients of the equations over the grid points, their performance will deteriorate as the variations in the coefficients increase. We therefore first symmetrically scale A by its diagonal before applying the BCCB preconditioners. In our tests, we apply diagonal scaling to all methods. We note that $O(m^2 \log m)$ flops are required by using the BCCB preconditioners, which is slightly more expensive than $O(m^2)$ flops required by the MILU preconditioners. The FFTs, however, can be computed in $O(\log m)$ parallel steps with $O(m^2)$ processors, see [90], whereas the MILU preconditioners require at least $O(m)$ steps regardless of how many processors could be used.

Tables 9.4–9.9 show the number of iterations required for convergence for different choices of ϵ and α. In the tables, I, C and M represent the systems with no preconditioner, BCCB preconditioner and MILU preconditioner respectively. We see that for small values of ϵ (e.g. $\epsilon \leq 0.01$) and large values of α (e.g. $\alpha \geq 100$), the number of iterations of the BCCB preconditioners is less than that of the MILU preconditioners. We also note that the MILU preconditioner is less sensitive to the changes in ϵ but more sensitive to the changes in α. In contrast, the BCCB preconditioner is less sensitive to the changes in α when α is large. In all cases, the number of iterations grows slower than as predicted by Theorems 9.2, 9.3 and 9.4.

Table 9.1. Eigenvalues distribution for $m = 4$

	A			$C^{-1}A$		
α	λ_1	λ_{m^2-1}	λ_{m^2}	λ_1	λ_{m^2-1}	λ_{m^2}
10	0.80393	6.2761	7.2761	0.80923	1.8355	7.0293
100	0.76433	6.2365	7.2365	0.80529	1.8460	8.3609
1000	0.76394	6.2361	7.2361	0.80525	1.8462	8.3775

Table 9.2. Eigenvalues distribution for $m = 8$

	A			$C^{-1}A$		
α	λ_1	λ_{m^2-1}	λ_{m^2}	λ_1	λ_{m^2-1}	λ_{m^2}
10	0.28123	7.4515	7.7988	0.64169	2.4046	9.1196
100	0.24373	7.4140	7.7613	0.63434	2.4791	16.896
1000	0.24123	7.4115	7.7588	0.63427	2.4798	17.040

Table 9.3. Eigenvalues distribution for $m = 16$

	A			$C^{-1}A$		
α	λ_1	λ_{m^2-1}	λ_{m^2}	λ_1	λ_{m^2-1}	λ_{m^2}
10	0.10811	7.8709	7.9719	0.57667	3.5551	8.5180
100	0.07061	7.8334	7.9344	0.56344	4.0515	32.934
1000	0.068112	7.8309	7.9319	0.56329	4.0577	34.342

Table 9.4. Number of iterations for (9.31) with $\alpha = 10$

ε	0.0			0.01			0.1			1.0		
m	I	C	M	I	C	M	I	C	M	I	C	M
8	24	12	10	24	14	10	28	14	10	29	15	10
16	44	16	13	47	18	13	50	18	13	53	20	13
32	72	19	15	72	22	15	78	22	15	89	25	16
64	94	26	15	94	30	15	103	30	15	120	33	17
128	107	37	15	107	43	15	113	44	15	139	47	17

Table 9.5. Number of iterations for (9.31) with $\alpha = 100$

ε	0.0			0.01			0.1			1.0		
m	I	C	M	I	C	M	I	C	M	I	C	M
8	24	12	11	25	13	11	29	14	10	29	15	10
16	47	16	15	53	18	15	54	18	15	57	20	14
32	89	19	21	102	22	21	103	23	21	109	26	20
64	171	25	30	198	29	30	201	30	30	213	33	29
128	326	32	40	351	38	40	356	40	40	416	45	40

Table 9.6. Number of iterations for (9.31) with $\alpha = 1000$

ε	0.0			0.01			0.1			1.0		
m	I	C	M	I	C	M	I	C	M	I	C	M
8	24	12	11	25	13	11	29	14	10	29	15	10
16	47	16	15	53	18	15	54	18	15	57	20	14
32	89	19	21	103	22	21	103	23	21	110	26	20
64	173	25	31	201	29	31	202	30	31	215	34	30
128	336	32	45	367	38	45	403	40	45	430	46	43

Table 9.7. Number of iterations for (9.32) with $\alpha = 10$

ε	0.0			0.01			0.1			1.0		
m	I	C	M	I	C	M	I	C	M	I	C	M
8	24	13	10	24	13	10	27	14	10	28	14	11
16	45	16	14	46	18	14	52	18	14	52	20	14
32	80	19	17	80	22	17	86	22	16	95	26	18
64	117	25	18	118	29	18	121	30	18	147	36	20
128	142	35	18	142	41	18	153	41	18	182	50	20

Table 9.8. Number of iterations for (9.32) with $\alpha = 100$

ε	0.0			0.01			0.1			1.0		
m	I	C	M	I	C	M	I	C	M	I	C	M
8	24	12	11	24	13	11	27	13	10	29	14	11
16	47	16	15	49	18	15	54	18	15	55	20	15
32	89	19	21	95	22	21	104	23	21	108	27	22
64	172	25	30	184	29	30	191	30	31	211	36	31
128	332	32	42	353	39	42	354	40	42	418	50	43

Table 9.9. Number of iterations for (9.32) with $\alpha = 1000$

ε	0.0			0.01			0.1			1.0		
m	I	C	M	I	C	M	I	C	M	I	C	M
8	24	12	11	24	13	11	27	13	10	29	14	11
16	47	16	15	49	18	15	54	18	15	55	20	16
32	89	19	21	95	22	21	104	23	21	108	27	22
64	173	25	31	186	29	31	192	30	31	213	36	31
128	336	32	45	359	38	45	371	40	45	427	51	45

9.5   3-dimensional problems

In this section we propose a preconditioner based on the fast sine transform for discretized systems of second-order partial differential equations in 3-dimensional space. The condition number of the preconditioned system is further reduced to $O(1)$. We will only consider elliptic equations. The extension to hyperbolic equations and parabolic equations is straightforward, see [51].

9.5.1   Discretized system

Let us consider the following second-order elliptic equation

$$\frac{\partial}{\partial x_1}\left(a^{(1)}\frac{\partial z}{\partial x_1}\right) + \frac{\partial}{\partial x_2}\left(a^{(2)}\frac{\partial z}{\partial x_2}\right) + \frac{\partial}{\partial x_3}\left(a^{(3)}\frac{\partial z}{\partial x_3}\right) = g \qquad (9.33)$$

where $0 < x_1,\ x_2,\ x_3 < 1$ and

$$a^{(1)} = a^{(1)}(x_1, x_2, x_3), \qquad a^{(2)} = a^{(2)}(x_1, x_2, x_3),$$

$$a^{(3)} = a^{(3)}(x_1, x_2, x_3), \qquad g = g(x_1, x_2, x_3),$$

are given functions with

$$0 < c_{\min} \le a^{(1)}(x_1, x_2, x_3),\ a^{(2)}(x_1, x_2, x_3),\ a^{(3)}(x_1, x_2, x_3) \le c_{\max} \quad (9.34)$$

for some constants $c_{\min}$ and $c_{\max}$. The boundary conditions are given by

$$z(0, x_2, x_3) = z_1(x_2, x_3), \qquad z(1, x_2, x_3) = z_2(x_2, x_3),$$

$$z(x_1, 0, x_3) = z_3(x_1, x_3), \qquad z(x_1, 1, x_3) = z_4(x_1, x_3),$$

$$z(x_1, x_2, 0) = z_5(x_1, x_2), \qquad z(x_1, x_2, 1) = z_6(x_1, x_2).$$

Uniform grid is imposed in the computational domain with $(m + 2)^3$ gridpoints, where $m \ge 2$. Let $z_{i,j,k}$ denote the calculated approximate solution of z at point $(x_{1,i}, x_{2,j}, x_{3,k})$ and $a^{(1)}_{i,j,k}, a^{(2)}_{i,j,k}, a^{(3)}_{i,j,k}, g_{i,j,k}$ denote the values of $a^{(1)}, a^{(2)}, a^{(3)}, g$ at point $(x_{1,i}, x_{2,j}, x_{3,k})$ respectively, where

$$\begin{cases} x_{1,i} = ih, & i = 0, \ldots, m+1, \\ x_{2,j} = jh, & j = 0, \ldots, m+1, \\ x_{3,k} = kh, & k = 0, \ldots, m+1, \end{cases}$$

and h is a space step.

Substitute the following differences into (9.33),

$$
\begin{cases}
\dfrac{\partial}{\partial x_1}\left(a^{(1)}\dfrac{\partial z}{\partial x_1}\right) \approx \dfrac{a^{(1)}_{i+\frac{1}{2},j,k}(z_{i+1,j,k}-z_{i,j,k}) - a^{(1)}_{i-\frac{1}{2},j,k}(z_{i,j,k}-z_{i-1,j,k})}{h^2}, \\[4mm]
\dfrac{\partial}{\partial x_2}\left(a^{(2)}\dfrac{\partial z}{\partial x_2}\right) \approx \dfrac{a^{(2)}_{i,j+\frac{1}{2},k}(z_{i,j+1,k}-z_{i,j,k}) - a^{(2)}_{i,j-\frac{1}{2},k}(z_{i,j,k}-z_{i,j-1,k})}{h^2}, \\[4mm]
\dfrac{\partial}{\partial x_3}\left(a^{(3)}\dfrac{\partial z}{\partial x_3}\right) \approx \dfrac{a^{(3)}_{i,j,k+\frac{1}{2}}(z_{i,j,k+1}-z_{i,j,k}) - a^{(3)}_{i,j,k-\frac{1}{2}}(z_{i,j,k}-z_{i,j,k-1})}{h^2},
\end{cases}
$$

we then have

$$
\left(a^{(1)}_{i+\frac{1}{2},j,k} + a^{(1)}_{i-\frac{1}{2},j,k} + a^{(2)}_{i,j+\frac{1}{2},k} + a^{(2)}_{i,j-\frac{1}{2},k} + a^{(3)}_{i,j,k+\frac{1}{2}} + a^{(3)}_{i,j,k-\frac{1}{2}}\right)z_{i,j,k}
$$

$$
-a^{(1)}_{i+\frac{1}{2},j,k}z_{i+1,j,k} - a^{(1)}_{i-\frac{1}{2},j,k}z_{i-1,j,k} - a^{(2)}_{i,j+\frac{1}{2},k}z_{i,j+1,k} - a^{(2)}_{i,j-\frac{1}{2},k}z_{i,j-1,k}
$$

$$
-a^{(3)}_{i,j,k+\frac{1}{2}}z_{i,j,k+1} - a^{(3)}_{i,j,k-\frac{1}{2}}z_{i,j,k-1} = h^2 g_{i,j,k}.
$$

Let

$$
z \equiv (z_{1,1,1}, \cdots, z_{m,1,1}, z_{1,2,1}, \cdots, z_{m,2,1}, \cdots, z_{1,m,m}, \cdots, z_{m,m,m})^T
$$

and

$$
d \equiv h^2 (g_{1,1,1}, \cdots, g_{m,1,1}, g_{1,2,1}, \cdots, g_{m,2,1}, \cdots, g_{1,m,m}, \cdots, g_{m,m,m})^T,
$$

we then have the following linear system

$$
Az = d \tag{9.35}
$$

where $A \in \mathbb{R}^{m^3 \times m^3}$.

9.5.2 Preconditioner based on sine transform

We define a preconditioner based on the fast sine transform for the discretized system (9.35). Let

$$
\begin{cases}
\bar{a}_1 = \dfrac{1}{m^2(m-1)} \sum_{k=1}^{m} \sum_{j=1}^{m} \sum_{i=1}^{m-1} a^{(1)}(x_{1,i+\frac{1}{2}}, x_{2,j}, x_{3,k}), \\[2mm]
\bar{a}_2 = \dfrac{1}{m^2(m-1)} \sum_{k=1}^{m} \sum_{j=1}^{m-1} \sum_{i=1}^{m} a^{(2)}(x_{1,i}, x_{2,j+\frac{1}{2}}, x_{3,k}), \\[2mm]
\bar{a}_3 = \dfrac{1}{m^2(m-1)} \sum_{k=1}^{m-1} \sum_{j=1}^{m} \sum_{i=1}^{m} a^{(3)}(x_{1,i}, x_{2,j}, x_{3,k+\frac{1}{2}}).
\end{cases}
$$

The constants $\bar{a}_1$, $\bar{a}_2$ and $\bar{a}_3$ are averages of $a^{(1)}(x_1, x_2, x_3)$, $a^{(2)}(x_1, x_2, x_3)$ and $a^{(3)}(x_1, x_2, x_3)$ over the grid accordingly.

The preconditioner is of the following form,

$$
S = S_3 \otimes I \otimes I + I \otimes S_2 \otimes I + I \otimes I \otimes S_1 \tag{9.36}
$$

where S_1, S_2, S_3 are defined by

$$
S_1 = \bar{a}_1 A_0, \quad S_2 = \bar{a}_2 A_0, \quad S_3 = \bar{a}_3 A_0
$$

with

$$
A_0 \equiv
\begin{pmatrix}
2 & -1 & & 0 \\
-1 & 2 & \ddots & \\
& \ddots & \ddots & -1 \\
0 & & -1 & 2
\end{pmatrix}. \tag{9.37}
$$

We note that the system $Sx = b$ can be solved in $O(m^3 \log m)$ operations by using the 3-dimensional fast sine transform, see [25] and [51].

Now, let us analyse the condition number of the preconditioned system. Let $A \in \mathbb{R}^{m^3 \times m^3}$ be defined as in (9.35). In the following, we assume that the inequality (9.34) is satisfied in the unit cube. Let

$$
A_{\min} = c_{\min}(A_0 \otimes I \otimes I + I \otimes A_0 \otimes I + I \otimes I \otimes A_0)
$$

and

$$A_{\max} = c_{\max}(A_0 \otimes I \otimes I + I \otimes A_0 \otimes I + I \otimes I \otimes A_0),$$

where A_0 is given by (9.37). It is easy to prove that $A - A_{\min}$ and $A_{\max} - A$ are positive semi-definite. Since both

$$S_i - c_{\min} A_0 = (\bar{a}_i - c_{\min}) A_0$$

and

$$c_{\max} A_0 - S_i = (c_{\max} - \bar{a}_i) A_0$$

are positive semi-definite, for $i = 1, 2, 3$, therefore the differences

$$S - A_{\min} = (S_3 - c_{\min} A_0) \otimes I \otimes I + I \otimes (S_2 - c_{\min} A_0) \otimes I + I \otimes I \otimes (S_1 - c_{\min} A_0)$$

and

$$A_{\max} - S = (c_{\max} A_0 - S_3) \otimes I \otimes I + I \otimes (c_{\max} A_0 - S_2) \otimes I + I \otimes I \otimes (c_{\max} A_0 - S_1)$$

are positive semi-definite too. Thus, we get

$$0 < \frac{c_{\min}}{c_{\max}} = \frac{x^T A_{\min} x}{x^T A_{\max} x} \leq \frac{x^T A x}{x^T S x} \leq \frac{x^T A_{\max} x}{x^T A_{\min} x} = \frac{c_{\max}}{c_{\min}}.$$

We then have

Theorem 9.5 *For the condition number of preconditioned system corresponding to the 3-dimensional elliptic problem (9.33) under the condition (9.34) with the preconditioner S defined by (9.36), we have*

$$\kappa(S^{-1} A) \leq O(1).$$

We remark that from the condition number viewpoint, it seems that the preconditioner based on the fast sine transform is better than the BCCB preconditioner. However, if higher order centered difference schemes are used, the matrix A_0 is not a tri-diagonal matrix anymore. In this case, the preconditioner based on the fast sine transform is not available but we can still use the BCCB preconditioner.

Chapter 10

Applications in First-Order PDEs

In this chapter we propose the generalized minimal residual (GMRES) method [38, 75] with preconditioners based on the fast sine transform for solving non-symmetric and highly non-diagonally dominant linear systems that arise from discretizations of first-order partial differential equations. When the GMRES method is applied to solving preconditioned systems, we show that the asymptotic convergence factor of the method is independent of the mesh size and depends only on the grid ratio. Furthermore, we compare the convergence properties of our preconditioned system with the one preconditioned by the semi-Toeplitz preconditioner proposed in [44]. We show that our preconditioned system has a smaller asymptotic convergence factor and numerical experiments indicate that our preconditioned system has a much faster convergence rate. For a literature on iterative solvers for first-order partial differential equations, we refer to [44, 45, 46, 47, 48, 49].

10.1 Discretized system and GMRES method

In this section we derive discretized linear systems of first-order partial differential equations. Let us consider

$$\frac{\partial u}{\partial t} + \sigma_1(x)\frac{\partial u}{\partial x} + \sigma_2(y)\frac{\partial u}{\partial y} = g(x,y,t) \tag{10.1}$$

where $0 < x < 1$, $0 < y < 1$, $t > 0$, and $\sigma_1(x)$, $\sigma_2(y)$ are given functions with $\sigma_1(x)$, $\sigma_2(y) > 0$. The equation (10.1) is well-posed if we define the following initial condition

$$u(x,y,0) = f_2(x,y),$$

and boundary conditions

$$u(x,0,t) = f_0(x,t), \qquad u(0,y,t) = f_1(y,t),$$

suitably. The uniform grid is defined as

$$x_i = ih_1, \qquad y_j = jh_2,$$

for $i = 0, \cdots, m_1$ and $j = 0, \cdots, m_2$, where h_1 and h_2 are space steps in x and y directions respectively. If we use the trapezoidal rule to discretize the time with a step τ, the centered difference to discretize the interior spatial derivative and the backward difference at boundaries $(x,1,t)$ and $(1,y,t)$, then we have the following linear systems of equations:

$$\frac{u_{i,j}^{n+1} - u_{i,j}^{n}}{\tau} + \sigma_1(x_i)\frac{u_{i+1,j}^{n+1} - u_{i-1,j}^{n+1} + u_{i+1,j}^{n} - u_{i-1,j}^{n}}{4h_1}$$

$$+\sigma_2(y_j)\frac{u_{i,j+1}^{n+1} - u_{i,j-1}^{n+1} + u_{i,j+1}^{n} - u_{i,j-1}^{n}}{4h_2} = g_{i,j}^{n},$$

for $i = 1, \cdots, m_1 - 1$ and $j = 1, \cdots, m_2 - 1$;

$$\frac{u_{m_1,j}^{n+1} - u_{m_1,j}^{n}}{\tau} + \sigma_1(x_{m_1})\frac{u_{m_1,j}^{n+1} - u_{m_1-1,j}^{n+1} + u_{m_1,j}^{n} - u_{m_1-1,j}^{n}}{2h_1}$$

$$+\sigma_2(y_j)\frac{u_{m_1,j+1}^{n+1} - u_{m_1,j-1}^{n+1} + u_{m_1,j+1}^{n} - u_{m_1,j-1}^{n}}{4h_2} = g_{m_1,j}^{n},$$

for $j = 1, \cdots, m_2 - 1$;

$$\frac{u_{i,m_2}^{n+1} - u_{i,m_2}^n}{\tau} + \sigma_1(x_i)\frac{u_{i+1,m_2}^{n+1} - u_{i-1,m_2}^{n+1} + u_{i+1,m_2}^n - u_{i-1,m_2}^n}{4h_1}$$

$$+\sigma_2(y_{m_2})\frac{u_{i,m_2}^{n+1} - u_{i,m_2-1}^{n+1} + u_{i,m_2}^n - u_{i,m_2-1}^n}{2h_2} = g_{i,m_2}^n,$$

for $i = 1, \cdots, m_1 - 1$; and

$$\frac{u_{m_1,m_2}^{n+1} - u_{m_1,m_2}^n}{\tau} + \sigma_1(x_{m_1})\frac{u_{m_1,m_2}^{n+1} - u_{m_1-1,m_2}^{n+1} + u_{m_1,m_2}^n - u_{m_1-1,m_2}^n}{2h_1}$$

$$+\sigma_2(y_{m_2})\frac{u_{m_1,m_2}^{n+1} - u_{m_1,m_2-1}^{n+1} + u_{m_1,m_2}^n - u_{m_1,m_2-1}^n}{2h_2} = g_{m_1,m_2}^n.$$

Let

$$u^n \equiv (u_{1,1}^n, u_{2,1}^n, \cdots, u_{m_1,1}^n, u_{1,2}^n, \cdots, u_{1,m_2}^n, \cdots, u_{m_1,m_2}^n)^T,$$

$$g^n \equiv (g_{1,1}^n, g_{2,1}^n, \cdots, g_{m_1,1}^n, g_{1,2}^n, \cdots, g_{1,m_2}^n, \cdots, g_{m_1,m_2}^n)^T,$$

$$\kappa_1 = \tau/h_1, \qquad \kappa_2 = \tau/h_2$$

and

$$\tilde{\kappa}_{1,i} = \sigma_1(x_i)\kappa_1, \qquad \tilde{\kappa}_{2,j} = \sigma_2(y_j)\kappa_2,$$

for $i = 1, \cdots, m_1$ and $j = 1, \cdots, m_2$. The linear system can now be expressed as

$$Au^{n+1} = b^{n+1}$$

where

$$A = \begin{pmatrix} H & G_{(1)} & & & \\ -G_{(2)} & H & G_{(2)} & & \\ & \ddots & \ddots & \ddots & \\ & & -G_{(m_2-1)} & H & G_{(m_2-1)} \\ & & & -2G_{(m_2)} & H + 2G_{(m_2)} \end{pmatrix} \in \mathbb{R}^{m_1 m_2 \times m_1 m_2}$$

with

$$H = \begin{pmatrix} 4 & \tilde{\kappa}_{1,1} & & & \\ -\tilde{\kappa}_{1,2} & \ddots & \tilde{\kappa}_{1,2} & & \\ & \ddots & \ddots & \ddots & \\ & & -\tilde{\kappa}_{1,m_1-1} & 4 & \tilde{\kappa}_{1,m_1-1} \\ & & & -2\tilde{\kappa}_{1,m_1} & 4 + 2\tilde{\kappa}_{1,m_1} \end{pmatrix}$$

and

$$G_{(j)} = \tilde{\kappa}_{2,j} I_{m_1},$$

for $j = 1, \cdots, m_2$. The vector b^{n+1} is known and depends only on u^n and g^n. We remark that the matrix A can also be expressed as a tensor form:

$$A = 4I_{m_1 m_2} + I_{m_2} \otimes A_{(1)} + A_{(2)} \otimes I_{m_1} \tag{10.2}$$

where

$$A_{(1)} = \begin{pmatrix} 0 & \tilde{\kappa}_{1,1} & & & \\ -\tilde{\kappa}_{1,2} & \ddots & \tilde{\kappa}_{1,2} & & \\ & \ddots & \ddots & \ddots & \\ & & -\tilde{\kappa}_{1,m_1-1} & 0 & \tilde{\kappa}_{1,m_1-1} \\ & & & -2\tilde{\kappa}_{1,m_1} & 2\tilde{\kappa}_{1,m_1} \end{pmatrix} \in \mathbb{R}^{m_1 \times m_1}$$

and

$$A_{(2)} = \begin{pmatrix} 0 & \tilde{\kappa}_{2,1} & & & \\ -\tilde{\kappa}_{2,2} & \ddots & \tilde{\kappa}_{2,2} & & \\ & \ddots & \ddots & \ddots & \\ & & -\tilde{\kappa}_{2,m_2-1} & 0 & \tilde{\kappa}_{2,m_2-1} \\ & & & -2\tilde{\kappa}_{2,m_2} & 2\tilde{\kappa}_{2,m_2} \end{pmatrix} \in \mathbb{R}^{m_2 \times m_2}.$$

The matrix A defined as in (10.2) is non-symmetric and usually not diagonally dominant. Classical stationary iterative methods, such as the Jacobi, Gauss–Seidel or SOR methods, may not converge. We will apply the GMRES method to solving the linear system.

For general linear systems $Ax = b$ where $A \in \mathbb{R}^{n \times n}$ is invertible, the scheme of the GMRES method is given as follows, see [38, 75]. At the initialization step, let

$$r^{(0)} = b - Ax^{(0)}, \quad \beta = \|r^{(0)}\|_2, \quad v^{(1)} = r^{(0)}/\beta.$$

In the iteration steps, we have

$$
\begin{cases}
\text{for } j = 1, 2, \cdots, m, \text{ do} \\[1.5ex]
\qquad w^{(j)} := Av^{(j)}, \\[1.5ex]
\qquad \text{for } i = 1, \cdots, j, \text{ do} \\[1.5ex]
\qquad\qquad h_{ij} := \langle w^{(j)}, v^{(i)} \rangle, \\[1.5ex]
\qquad\qquad w^{(j)} := w^{(j)} - h_{ij}v^{(i)}, \\[1.5ex]
\qquad \text{end do} \\[1.5ex]
\qquad h_{j+1,j} := \|w^{(j)}\|_2; \ \text{if } h_{j+1,j} = 0, \text{ set } m := j \text{ and go to } (*), \\[1.5ex]
\qquad v^{(j+1)} := w^{(j+1)}/h_{j+1,j}, \\[1.5ex]
\text{end do} \\[1.5ex]
(*) \text{ compute } y^{(m)} \text{ the minimizer of } \|\beta e_1 - \bar{H}_m y\|_2, \\[1.5ex]
x^{(m)} := x^{(0)} + V_m y^{(m)},
\end{cases}
$$

where $\bar{H}_m = (h_{ij})_{1 \le i \le m+1, 1 \le j \le m}$ and $V_m \in \mathbb{R}^{n \times m}$ is a matrix with column vectors $v^{(1)}, \cdots, v^{(m)}$.

In the j-th iteration the GMRES method generates a residual vector $r^{(j)}$ that satisfies

$$
\|r^{(j)}\|_2 = \min_{p_j \in P_j, p_j(0) = 1} \|p_j(M^{-1}A)r^{(0)}\|_2,
$$

where P_j is the set of all polynomials of degree j and M is a preconditioner. In addition, if $M^{-1}A$ is diagonalizable, then we have

$$
\frac{\|r^{(j)}\|_2}{\|r^{(0)}\|_2} \le \kappa(Q_{M^{-1}A}) \cdot \min_{p_j \in P_j, p_j(0) = 1} \max_k |p_j(\lambda_k)| \tag{10.3}
$$

where $Q_{M^{-1}A}$ is the eigenvector matrix of $M^{-1}A$, $\kappa(Q_{M^{-1}A})$ denotes the condition number of $Q_{M^{-1}A}$ and λ_k are the eigenvalues of $M^{-1}A$, see [75].

Therefore if $M^{-1}A$ has N distinct eigenvalues the GMRES method will converge to the true solution within $N+1$ iterations in exact arithmetric.

10.2 Construction of preconditioner

We are going to construct a preconditioner for A defined as in (10.2). In [44], a semi-Toeplitz preconditioner T was constructed for A. More precisely, T was obtained by using a Toeplitz matrix to approximate $A_{(1)}$. We will follow the same approach but we make use of a near-by Toeplitz matrix to approximate $A_{(1)}$ instead. Our preconditioner M is defined as follows:

$$M = 4I_{m_1 m_2} + \alpha I_{m_2} \otimes \tilde{R}_{m_1} + A_{(2)} \otimes I_{m_1}$$

where

$$\alpha = \frac{1}{2(m_1 - 1)} \sum_{j=1}^{m_1 - 1} (\tilde{\kappa}_{1,j} + \tilde{\kappa}_{1,j+1})$$

and

$$\tilde{R}_{m_1} \equiv \begin{pmatrix} 0 & 1 & & & \\ -1 & \ddots & 1 & & \\ & \ddots & \ddots & \ddots & \\ & & -1 & 0 & 1 \\ & & & -2 & 0 \end{pmatrix} \in \mathbb{R}^{m_1 \times m_1}.$$

In the following we present a fast algorithm for computing the matrix-vector multiplication $M^{-1}v$.

Theorem 10.1 The matrix $\tilde{R}_{m_1}$ can be diagonalized as follows,

$$S^{-1}\tilde{R}_{m_1} S = \mathrm{diag}(\lambda_{1,1}, \cdots, \lambda_{1,m_1})$$

where the entries of S are given by

$$(S)_{j,k} = i^j \sin \frac{(2k-1)j\pi}{2m_1}, \qquad i \equiv \sqrt{-1}, \tag{10.4}$$

for $j, k = 1, \cdots, m_1$, and

$$\lambda_{1,j} = 2i \cos \frac{(2j-1)\pi}{2m_1}, \tag{10.5}$$

for $j = 1, \cdots, m_1$.

Proof: By direct verification, we know that the j-th column of S is an eigenvector of $\tilde{R}_{m_1}$ and $\lambda_{1,j}$ is its corresponding eigenvalue. $\square$

By implementing a modified version of the discrete forward sine transform similarly to [44], the matrix-vector multiplications Sv and S^*v can be computed in $O(m_1 \log m_1)$ operations. The following theorem will help us to compute $S^{-1}v$ fast. The proof of the theorem is straightforward.

Theorem 10.2 *We have*

$$S^{-1} = S^*\mathrm{diag}\Big(\frac{2}{m_1},\cdots,\frac{2}{m_1},\frac{1}{m_1}\Big). \tag{10.6}$$

With the help of (10.6) and the modified discrete forward sine transform, $S^{-1}v$ can also be done in $O(m_1 \log m_1)$ operations. Now, our preconditioner M can be decomposed as

$$M = (I_{m_2} \otimes S)\Lambda(I_{m_2} \otimes S^{-1}),$$

where Λ is a block tri-diagonal matrix with diagonal blocks. Therefore, the multiplication $M^{-1}v$ will require $2m_2$ modified discrete forward sine transforms, see [44], and m_1 tri-diagonal solvers, each of order m_2. Hence, the matrix-vector multiplication $M^{-1}v$ costs $O(m_1 m_2 \log m_1)$ operations, see [44, 48, 49].

10.3   Convergence rate

Let $\sigma_1(x) \equiv \sigma_2(y) \equiv 1$. We are going to analyse the convergence rate of the GMRES method. In this constant-coefficient case, the matrices A and M can be written as

$$A = 4I_{m_1 m_2} + \kappa_1 I_{m_2} \otimes R_{m_1} + \kappa_2 R_{m_2} \otimes I_{m_1}$$

and

$$M = 4I_{m_1 m_2} + \kappa_1 I_{m_2} \otimes \tilde{R}_{m_1} + \kappa_2 R_{m_2} \otimes I_{m_1}$$

where

$$R_{m_j} \equiv \begin{pmatrix} 0 & 1 & & & \\ -1 & \ddots & 1 & & \\ & \ddots & \ddots & \ddots & \\ & & -1 & 0 & 1 \\ & & & -2 & 2 \end{pmatrix} \in \mathbb{R}^{m_j \times m_j} \tag{10.7}$$

for $j = 1, 2$. Since R_{m_1} and R_{m_2} have the same structure, we will only consider R_{m_2} in the later discussion.

Let
$$E \equiv A - M.$$

Then $E = I_{m_2} \otimes E_{(1)}$ where $E_{(1)} \in \mathbb{R}^{m_1 \times m_1}$ is a rank one matrix given by

$$E_{(1)} = \begin{pmatrix} 0 \cdots 0 & 0 \\ \vdots & \vdots & \vdots \\ 0 \ldots 0 & 0 \\ 0 \ldots 0 & 2\kappa_1 \end{pmatrix} = 2\kappa_1 e_{m_1} e_{m_1}^T \qquad (10.8)$$

where $e_{m_1} \in \mathbb{R}^{m_1}$ is the m_1-th unit vector. Therefore,

$$\mathrm{rank}(M^{-1}A - I_{m_1 m_2}) = \mathrm{rank}(M^{-1}(I_{m_2} \otimes E_{(1)})) = m_2.$$

In order to analyse the convergence rate of our method, we now review some well known facts about Krylov subspace methods.

For solving a linear system $Gu = b$, Krylov subspace methods are a class of methods which seek an approximation $u^{(m)}$ from an affine subspace $u^{(0)} + \mathcal{K}_m(G, r^{(0)})$, where $u^{(0)}$ is the initial guess to the solution, $r^{(0)}$ is the initial residual and

$$\mathcal{K}_m(G, r^{(0)}) \equiv \mathrm{span}\{r^{(0)}, Gr^{(0)}, G^2 r^{(0)}, \cdots, G^{m-1} r^{(0)}\}.$$

We have

Lemma 10.1 *When Krylov subspace methods are applied to solving a linear system $Gu = b$ where $G = I + L$ and I is the identity matrix, the methods will converge in at most $\mathrm{rank}(L)+1$ iterations in exact arithmetic.*

Proof: We first recall that the minimal polynomial of $r^{(0)}$ with respect to G is the non-zero monic polynomial p of the lowest degree such that $p(G)r^{(0)} = 0$. Krylov subspace methods must converge with ν steps, where ν is the degree of the minimal polynomials of the residual $r^{(0)}$ with respect to $G = I + L$, see [75]. Let μ be the degree of the minimal polynomials of $r^{(0)}$ with respect to L. Since

$$\sum_{i=0}^{k} \alpha_i (I + L)^i r^{(0)} = 0$$

implies

$$\sum_{i=0}^{k} \beta_i L^i r^{(0)} = 0$$

for some constants β_i and vice versa, we have $\nu = \mu$. Moreover, from the definition of μ, the set

$$\{r^{(0)}, Lr^{(0)}, \cdots, L^{\mu-1}r^{(0)}\}$$

is linearly independent. Let B be the column vector space of L. Then, the dimension of B is equal to the rank of L. Since $L^i r^{(0)} \in B$ for $i \geq 1$, we have

$$\{Lr^{(0)}, \cdots, L^{\mu-1}r^{(0)}\} \subseteq B.$$

Thus, $\mu - 1 \leq \operatorname{rank}(L)$, i.e., $\mu \leq \operatorname{rank}(L) + 1$. Hence, Krylov subspace methods converge within

$$\nu = \mu \leq \operatorname{rank}(L) + 1$$

steps. □

Since the GMRES method is one of Krylov subspace methods, we have by using Lemma 10.1,

Theorem 10.3 *Let $\sigma_1(x) \equiv \sigma_2(y) \equiv 1$. The GMRES method converges in at most $m_2 + 1$ iterations in exact arithmetic when it is applied to solving the preconditioned system*

$$M^{-1}Au^{n+1} = M^{-1}b^{n+1}.$$

We then expect that the GMRES method would converge fast if m_2 is not very large.

10.4   Spectral analysis

We are going to study the spectrum of the preconditioned matrix $M^{-1}A$ by devising formulae for its eigenvalues. We need a lemma proposed in [44] for the spectrum of R_{m_2} defined by (10.7). Let $\mathcal{E}(a, b) \subset \mathbb{C}$ denote the closed ellipse centered at the origin with the semi-major axis b oriented along the imaginary axis and the semi-minor axis a, and let

$$\mathcal{E}^+(a, b) \equiv \{z \mid z \in \mathcal{E}(a, b) \text{ with real part } \operatorname{Re}(z) \geq 0\}.$$

Lemma 10.2 *The eigenvalues $\lambda_{2,k}$ of R_{m_2}, for $k = 1, \cdots, m_2$, satisfy:*

(i) $\lambda_{2,p} \neq \lambda_{2,q}$ *if $p \neq q$.*

(ii) $\lambda_{2,k} \in \mathcal{E}^+(4m_2^{-3/4}, 2 + 4m_2^{-3/2})$.

Note that

$$M^{-1}A = I + M^{-1}E.$$

In the following, let us compute the eigenvalues of $M^{-1}E$ first. From Lemma 10.2, we know that the matrix R_{m_2} is diagonalizable. Let V be the eigenvector matrix for R_{m_2}, i.e.,

$$V^{-1}R_{m_2}V = \text{diag}(\lambda_{2,1}, \cdots, \lambda_{2,m_1}). \tag{10.9}$$

Then, the matrix M can be diagonalized as follows,

$$M = (V \otimes S)D(V \otimes S)^{-1}$$

where S is defined by (10.4), and

$$D = \text{diag}(D_{(1)}, \cdots, D_{(m_2)})$$

with

$$D_{(k)} = \text{diag}(4 + \kappa_1\lambda_{1,1} + \kappa_2\lambda_{2,k}, \cdots, 4 + \kappa_1\lambda_{1,m_1} + \kappa_2\lambda_{2,k}),$$

for $k = 1, \cdots, m_2$. The $\lambda_{1,j}$ here, for $j = 1, \cdots, m_1$, are defined by (10.5). Now we have

$$(V \otimes I)^{-1}M^{-1}E(V \otimes I) = (I \otimes S)D^{-1}(I \otimes S^{-1}E_{(1)})$$

$$= \text{diag}(SD_{(1)}^{-1}S^{-1}E_{(1)}, \cdots, SD_{(m_2)}^{-1}S^{-1}E_{(1)}). \tag{10.10}$$

Let

$$W \equiv (V \otimes I)^{-1}M^{-1}E(V \otimes I).$$

Then W and $M^{-1}E$ have the same eigenvalues. Moreover,

$$W = \text{diag}(W_{(1)}, \cdots, W_{(m_2)})$$

with $W_{(k)} \equiv SD_{(k)}^{-1}S^{-1}E_{(1)}$. By (10.8), $W_{(k)}$ is of the following form:

$$
W_{(k)} = \begin{pmatrix} 0 \cdots 0 & W_{(k)}(1, m_1) \\ \vdots \quad \vdots & \vdots \\ 0 \ldots 0 & W_{(k)}(m_1 - 1, m_1) \\ 0 \ldots 0 & W_{(k)}(m_1, m_1) \end{pmatrix} \in \mathbb{R}^{m_1 \times m_1}. \tag{10.11}
$$

Obviously, the eigenvalues of $W_{(k)}$ are 0 with multiplicity $m_1 - 1$ and $W_{(k)}(m_1, m_1)$. Let us assume that $W_{(k)}(m_1, m_1) \neq 0$ as it will be proved later in Theorems 10.6 and 10.7. The following theorem identifies the last column of $W_{(k)}$.

Theorem 10.4 *Let $i \equiv \sqrt{-1}$. The entries of the last column of $W_{(k)}$ are given by*

$$
W_{(k)}(l, m_1) = \frac{2\kappa_1}{m_1} \sum_{j=1}^{m_1} \frac{i^l(-i)^{m_1}(-1)^{j+1} \sin \frac{(2j-1)l\pi}{2m_1}}{4 + \kappa_1\lambda_{1,j} + \kappa_2\lambda_{2,k}}, \tag{10.12}
$$

for $l = 1, \cdots, m_1$.

Proof: Note that

$$
W_{(k)} \equiv SD_{(k)}^{-1}S^{-1}E_{(1)}.
$$

By using (10.8) the last column of $W_{(k)}$ is given by

$$
2\kappa_1 SD_{(k)}^{-1}S^{-1}e_{m_1} = \frac{2\kappa_1}{m_1} SD_{(k)}^{-1}S^*e_{m_1}
$$

$$
= \frac{2\kappa_1}{m_1} SD_{(k)}^{-1}\left((-i)^{m_1} \sin \frac{(2j-1)m_1\pi}{2m_1}\right)_{j=1,\cdots,m_1}
$$

$$
= \frac{2\kappa_1}{m_1} SD_{(k)}^{-1}\left((-i)^{m_1}(-1)^{j+1}\right)_{j=1,\cdots,m_1}
$$

$$
= \frac{2\kappa_1}{m_1} S\left(\frac{(-i)^{m_1}(-1)^{j+1}}{4 + \kappa_1\lambda_{1,j} + \kappa_2\lambda_{2,k}}\right)_{j=1,\cdots,m_1}
$$

$$
= \left(\frac{2\kappa_1}{m_1} \sum_{j=1}^{m_1} \frac{i^l(-i)^{m_1}(-1)^{j+1} \sin \frac{(2j-1)l\pi}{2m_1}}{4 + \kappa_1\lambda_{1,j} + \kappa_2\lambda_{2,k}}\right)_{l=1,\cdots,m_1}
$$

where $(\alpha_p)_{p=1,\cdots,m_1}$ denotes the column vector $(\alpha_1,\cdots,\alpha_{m_1})^T$. $\square$

Therefore, the non-zero eigenvalue of $W_{(k)}$ can be obtained by putting $l = m_1$ in (10.12).

Corollary 10.1 *The non-zero eigenvalue of* $W_{(k)}$ *is given by*

$$\lambda_k \equiv W_{(k)}(m_1, m_1) = \frac{2\kappa_1}{m_1} \sum_{j=1}^{m_1} \frac{1}{4 + \kappa_1\lambda_{1,j} + \kappa_2\lambda_{2,k}}. \tag{10.13}$$

Thus, we need to compute the sum in (10.13) in order to find explicit formulae for λ_k, for $k = 1,\cdots,m_2$.

To compute the sum in (10.13), let us first introduce the following notations:

$$F_{k,l}(\theta) \equiv \frac{\sin(k\theta)\sin(l\theta)}{2\alpha + 2i\beta\cos\theta}, \qquad i \equiv \sqrt{-1}, \tag{10.14}$$

and

$$\Phi_{k,l}(\alpha,\beta) \equiv \frac{2}{m_1} \sum_{j=1}^{m_1} F_{k,l}\left(\frac{(2j-1)\pi}{2m_1}\right), \tag{10.15}$$

for $k,l = 1,\cdots,m_1$, where $\alpha \in \mathbb{C}$ with $\mathrm{Re}(\alpha) > 0$ and $\beta > 0$. Recall that in (10.5), we have

$$\lambda_{1,j} = 2i\cos\frac{(2j-1)\pi}{2m_1},$$

therefore λ_k can be expressed as

$$\lambda_k = \kappa_1\Phi_{m_1,m_1}\left(\frac{4+\kappa_2\lambda_{2,k}}{2}, \kappa_1\right). \tag{10.16}$$

We can devise an analytic expression for the sum in (10.15).

Lemma 10.3 *The* $\Phi_{k,l}(\alpha,\beta)$ *can be written as follows,*

$$\Phi_{k,l}(\alpha,\beta) = \frac{i}{\beta(z-z^{-1})}\left(z^{k+l} - z^{|k-l|} - (z^k - z^{-k})(z^l - z^{-l})\frac{z^{2m_1}}{1+z^{2m_1}}\right)$$

where

$$z = i\left(\frac{\alpha}{\beta} - \sqrt{1 + \left(\frac{\alpha}{\beta}\right)^2}\right)$$

with an assumption $|z| \neq 1$ *and '*$-\sqrt{}$*' denotes the complex square root such that* $|z| < 1$.

Proof: Since $F_{k,l}(\theta) \in C_{2\pi}^1$, we have

$$F_{k,l}(\theta) = \sum_{q=-\infty}^{+\infty} c_q e^{iq\theta} \qquad \text{and} \qquad c_q = \frac{1}{2\pi} \int_{-\pi}^{\pi} e^{-iq\theta} F_{k,l}(\theta) d\theta.$$

By using the Poisson summation formula, see [73] and [101], we have

$$\frac{1}{2m_1} \sum_{j=0}^{2m_1-1} F_{k,l}\left(\frac{2j\pi}{2m_1}\right) = \sum_{q=-\infty}^{+\infty} c_{2m_1 q}$$

and

$$\frac{1}{4m_1} \sum_{j=0}^{4m_1-1} F_{k,l}\left(\frac{2j\pi}{4m_1}\right) = \sum_{q=-\infty}^{+\infty} c_{4m_1 q}.$$

Therefore,

$$\Phi_{k,l}(\alpha, \beta) = \frac{2}{m_1} \sum_{j=1}^{m_1} F_{k,l}\left(\frac{(2j-1)\pi}{2m_1}\right)$$

$$= \frac{1}{m_1} \sum_{j=1}^{2m_1} F_{k,l}\left(\frac{(2j-1)\pi}{2m_1}\right)$$

$$= \frac{1}{m_1} \sum_{j=0}^{4m_1-1} F_{k,l}\left(\frac{j\pi}{2m_1}\right) - \frac{1}{m_1} \sum_{j=0}^{2m_1-1} F_{k,l}\left(\frac{2j\pi}{2m_1}\right)$$

$$= 4 \sum_{q=-\infty}^{+\infty} c_{4m_1 q} - 2 \sum_{q=-\infty}^{+\infty} c_{2m_1 q}$$

$$= 2 \sum_{q=-\infty}^{+\infty} (-1)^q c_{2m_1 q}, \tag{10.17}$$

for $k, l = 1, \cdots, m_1$. Now, we calculate $c_{2m_1 q}$. By using the well known Euler identities in (10.14), we get

$$F_{k,l}(\theta) = -\frac{1}{4} \frac{(e^{ik\theta} - e^{-ik\theta})(e^{il\theta} - e^{-il\theta})}{2\alpha + i\beta(e^{i\theta} + e^{-i\theta})}$$

$$= \frac{e^{i(k-l)\theta} + e^{i(l-k)\theta} - e^{i(k+l)\theta} - e^{-i(k+l)\theta}}{8\alpha + 4\beta i(e^{i\theta} + e^{-i\theta})}.$$

Therefore,

$$c_q = \frac{1}{2\pi} \int_{-\pi}^{\pi} e^{-iq\theta} F_{k,l}(\theta) d\theta$$

$$= -\frac{i}{4\beta} \frac{1}{2\pi i} \oint_S \frac{\omega^{k-l-q} + \omega^{l-k-q} - \omega^{k+l-q} - \omega^{-l-k-q}}{(\omega - z)(\omega - z^{-1})} d\omega$$

(10.18)

where $\omega = e^{i\theta}$, $S \subset \mathbb{C}$ is the positively oriented unit circle, and

$$z = i\left(\frac{\alpha}{\beta} - \sqrt{1 + \left(\frac{\alpha}{\beta}\right)^2}\right)$$

(10.19)

with $|z| < 1$. By using the Residue Theorem, see [73], we have from (10.18),

$$\begin{cases} c_0 = -\frac{i}{2\beta} \frac{z^{|k-l|} - z^{k+l}}{z - z^{-1}}, \\[4mm] c_{2m_1 q} = -\frac{i}{4\beta(z - z^{-1})}\left(z^{2m_1 q - k + l} + z^{2m_1 q - l + k}\right. \\[4mm] \qquad\qquad \left. - z^{2m_1 q - k - l} - z^{2m_1 q + k + l}\right), \quad \text{for} \quad q > 0, \end{cases}$$

where z is defined as in (10.19). Since $c_{2m_1 q} = c_{-2m_1 q}$, we have by (10.17),

$$\Phi_{k,l}(\alpha, \beta) = 2\left(c_0 + 2\sum_{q=1}^{+\infty}(-1)^q c_{2m_1 q}\right)$$

$$= 2\left(\frac{-i}{4\beta(z - z^{-1})}\right)\left[2z^{|k-l|} - 2z^{k+l}\right.$$

$$\left. + 2\sum_{q=1}^{+\infty}(-1)^q\left(z^{2m_1 q - k + l} + z^{2m_1 q - l + k} - z^{2m_1 q - k - l} - z^{2m_1 q + k + l}\right)\right]$$

$$= \frac{i}{\beta(z - z^{-1})}\left(z^{k+l} - z^{|k-l|} - (z^k - z^{-k})(z^l - z^{-l})\frac{z^{2m_1}}{1 + z^{2m_1}}\right). \quad \square$$

Now Lemma 10.3 helps us to evaluate the sum defined as in (10.13).

Theorem 10.5 *We have*

$$\lambda_k = -\frac{2i}{z_k - z_k^{-1}} \frac{1 - z_k^{2m_1}}{1 + z_k^{2m_1}}, \qquad i \equiv \sqrt{-1},$$

where

$$z_k = i\left(\frac{4 + \kappa_2\lambda_{2,k}}{2\kappa_1} - \sqrt{1 + \left(\frac{4 + \kappa_2\lambda_{2,k}}{2\kappa_1}\right)^2}\right) \qquad (10.20)$$

with an assumption $|z_k| \neq 1$ *and '$-\sqrt{}$' denotes the complex square root such that* $|z_k| < 1$, $k = 1, \cdots, m_2$.

Proof: We have by (10.16) and Lemma 10.3,

$$\lambda_k = \kappa_1\Phi_{m_1,m_1}\left(\frac{4 + \kappa_2\lambda_{2,k}}{2}, \kappa_1\right)$$

$$= \frac{i}{z_k - z_k^{-1}}\left(z_k^{2m_1} - 1 - \frac{z_k^{2m_1}}{1 + z_k^{2m_1}}(z_k^{m_1} - z_k^{-m_1})^2\right)$$

$$= \frac{i}{z_k - z_k^{-1}}\frac{z_k^{4m_1} - 1 - (z_k^{2m_1} - 1)^2}{1 + z_k^{2m_1}}$$

$$= -\frac{2i}{z_k - z_k^{-1}}\frac{1 - z_k^{2m_1}}{1 + z_k^{2m_1}}, \qquad k = 1, \cdots, m_2,$$

where z_k is defined as in (10.20). $\square$

Since

$$M^{-1}A = I + M^{-1}E,$$

and the matrices $M^{-1}E$ and W have the same eigenvalues, therefore, all the eigenvalues of $M^{-1}A$ are determined.

Corollary 10.2 *The preconditioned matrix* $M^{-1}A$ *has* $(m_1-1)m_2$ *eigenvalues that are equal to 1. The other* m_2 *eigenvalues* μ_k *are given by*

$$\mu_k = 1 - \frac{2i}{z_k - z_k^{-1}}\frac{1 - z_k^{2m_1}}{1 + z_k^{2m_1}} \qquad (10.21)$$

with z_k *defined as in (10.20).*

10.5   Asymptotic properties

In this section, we study the asymptotic properties of the spectrum of the preconditioned matrix $M^{-1}A$.

Theorem 10.6 *Assume that the grid ratio*

$$\tau \equiv ch_1^\alpha \qquad (10.22)$$

where $0 < \alpha < 1$ and $c > 0$. Let

$$m_1 = \lceil m_2(1 + 2m_2^{-3/2})/\phi \rceil \qquad (10.23)$$

where $0 < \phi < 1$ and $\lceil a/b \rceil$ denotes the closest integer greater than a/b. When m_1 is sufficiently large, all the m_2 eigenvalues of $M^{-1}A$ that are different from 1 lie around a curve $\mu(\gamma)$ given by

$$\mu(\gamma) \equiv 1 + \frac{1}{\sqrt{1 - \gamma^2}} \qquad (10.24)$$

where $-\phi \le \gamma \le \phi$.

Proof: By Corollary 10.2, we have

$$\mu_k = 1 - \frac{2i}{z_k - z_k^{-1}} \frac{1 - z_k^{2m_1}}{1 + z_k^{2m_1}}$$

where z_k is defined by (10.20) with $|z_k| < 1$. It follows that

$$z_k^{2m_1} \to 0 \quad \text{as} \quad m_1 \to \infty.$$

Let

$$\zeta_k \equiv \frac{4 + \kappa_2 \lambda_{2,k}}{2\kappa_1},$$

for $k = 1, \cdots, m_2$. By (10.22), we obtain

$$\zeta_k = \frac{2h_1}{ch_1^\alpha} + \frac{h_1}{2h_2}\lambda_{2,k} = 2c^{-1}m_1^{\alpha-1} + \omega_k$$

where

$$\omega_k = \frac{m_2}{2m_1}\lambda_{2,k}. \qquad (10.25)$$

By using Lemma 10.2 and (10.25), we have

$$\omega_k \in \mathcal{E}^+\Big(\frac{2m_2^{1/4}}{m_1}, \frac{m_2}{m_1}(1 + 2m_2^{-3/2})\Big),$$

for $k = 1, \cdots, m_2$. We note that by (10.23),

$$m_2 < m_1 \quad \text{and} \quad \frac{m_2}{m_1}(1 + 2m_2^{-3/2}) \leq \phi,$$

therefore,

$$\omega_k \in \mathcal{E}^+\Big(\frac{2m_2^{1/4}}{m_1}, \phi\Big) \subset \mathcal{E}^+(2m_1^{-3/4}, \phi).$$

Now, let

$$\epsilon = c^{-1}m_1^{\alpha-1} + \delta_k m_1^{-3/4}$$

with $0 \leq \delta_k \leq 1$. We then have

$$\zeta_k = 2\epsilon + \gamma_k i \tag{10.26}$$

where $i \equiv \sqrt{-1}$ and $-\phi \leq \gamma_k \leq \phi$. When $m_1 \gg 1$, then $0 < \epsilon \ll 1$. By combining (10.26) with (10.20), we obtain

$$z_k = i\Big(\zeta_k - \sqrt{1 + \zeta_k^2}\Big) = 2\epsilon i - \gamma_k - i\sqrt{1 + (2\epsilon + \gamma_k i)^2}.$$

Therefore, when m_1 is sufficiently large,

$$z_k \approx -\gamma_k - i\sqrt{1 - \gamma_k^2}$$

with $-\phi \leq \gamma_k \leq \phi$. Furthermore, we have

$$z_k - z_k^{-1} \approx -2i\sqrt{1 - \gamma_k^2}.$$

Thus, when m_1 is sufficiently large,

$$\mu_k \approx 1 + \frac{1}{\sqrt{1 - \gamma_k^2}}$$

where $-\phi \leq \gamma_k \leq \phi$. Hence, all the eigenvalues of $M^{-1}A$ that are different from 1 lie around the curve given by (10.24). $\quad\square$

For the case $\alpha = 1$, we have the following theorem.

Theorem 10.7 *Assume that $\kappa_1 \equiv \tau/h_1$ is a large fixed constant and*

$$m_1 = \lceil m_2(1 + 2m_2^{-3/2})/\phi \rceil$$

with $0 < \phi < 1$. When m_1 is sufficiently large, all the m_2 eigenvalues of $M^{-1}A$ that are different from 1 lie around a curve $\mu(\gamma)$ given by

$$\mu(\gamma) \equiv 1 + \frac{1}{\sqrt{1-\gamma^2}} - \frac{2\gamma i}{\kappa_1(1-\gamma^2)^{3/2}} + O(\kappa_1^{-2}) \qquad (10.27)$$

where $i \equiv \sqrt{-1}$ and $-\phi \le \gamma \le \phi$.

Proof: We have by (10.21),

$$\mu_k = 1 - \frac{2i}{z_k - z_k^{-1}} \frac{1 - z_k^{2m_1}}{1 + z_k^{2m_1}}.$$

Let

$$\zeta_k \equiv \frac{4 + \kappa_2 \lambda_{2,k}}{2\kappa_1},$$

for $k = 1, \cdots, m_2$. Therefore,

$$\zeta_k = 2\kappa_1^{-1} + \omega_k$$

where

$$\omega_k = \frac{m_2}{2m_1} \lambda_{2,k}.$$

From the proof of Theorem 10.6, we note that

$$\omega_k \in \mathcal{E}^+(2m_1^{-3/4}, \phi).$$

Hence, ω_k can be written as

$$\omega_k = 2\delta_k m_1^{-3/4} + \gamma_k i$$

where $i \equiv \sqrt{-1}$, $0 \le \delta_k \le 1$, and $-\phi \le \gamma_k \le \phi$. Therefore,

$$\zeta_k = 2\kappa_1^{-1} + 2\delta_k m_1^{-3/4} + \gamma_k i. \qquad (10.28)$$

By using (10.20) and (10.28), we have

$$z_k = i\left(\zeta_k - \sqrt{1 + \zeta_k^2}\right)$$
$$= 2\kappa_1^{-1} i + 2\delta_k m_1^{-3/4} i - \gamma_k - i\sqrt{1 + (2\kappa_1^{-1} + 2\delta_k m_1^{-3/4} + \gamma_k i)^2}.$$

When m_1 is sufficiently large, we obtain

$$z_k \approx -\gamma_k + 2\kappa_1^{-1}i - i\sqrt{1 + (2\kappa_1^{-1} + \gamma_k i)^2}$$

with $-\phi \leq \gamma_k \leq \phi$. Furthermore, when m_1 is sufficiently large, one can easily obtain

$$\mu_k \approx 1 + \frac{1}{\sqrt{1 + (2\kappa_1^{-1} + \gamma_k i)^2}}$$

where $-\phi \leq \gamma_k \leq \phi$. By using Taylor expansion, we have

$$1 + \frac{1}{\sqrt{1 + (2\kappa_1^{-1} + \gamma_k i)^2}} = 1 + \frac{1}{\sqrt{1 - \gamma_k^2}} - \frac{2\gamma_k i}{\kappa_1(1 - \gamma_k^2)^{3/2}} + O(\kappa_1^{-2}).$$

Hence, all the eigenvalues of $M^{-1}A$ that are different from 1 lie around the curve given by (10.27). □

In Figure 10.1, we show how well the asymptotic formula given by (10.24) agrees with the eigenvalues given by (10.21) as m_2 increases. We observe that the eigenvalues almost lie on the interval $[2, 2.67]$ as m_2 increases ($1 + \frac{1}{\sqrt{1-\phi^2}} \approx 2.67$ when $\phi = 0.8$). In Figure 10.2, we show a good fit of the asymptotic formula given by (10.27) for the eigenvalues given by (10.21).

10.6   Convergence results and numerical tests

We compare the asymptotic convergence factor and the condition number reduction of our preconditioner with those of the semi-Toeplitz preconditioner proposed in [44]. The semi-Toeplitz preconditioner T is defined as follows:

$$T = 4I_{m_1 m_2} + \kappa_1 I_{m_2} \otimes \hat{R}_{m_1} + \kappa_2 R_{m_2} \otimes I_{m_1} \tag{10.29}$$

where

$$\hat{R}_{m_1} \equiv \begin{pmatrix} 0 & 1 & & & \\ -1 & \ddots & 1 & & \\ & \ddots & \ddots & \ddots & \\ & & -1 & 0 & 1 \\ & & & -1 & 0 \end{pmatrix} \in \mathbb{R}^{m_1 \times m_1}$$

and R_{m_2} is defined as in (10.7).

Now, we introduce the asymptotic convergence factor ρ which is defined as

$$\rho \equiv \lim_{j \to \infty} \left(\min_{p_j \in P_j, p_j(0)=1} \max_k |p_j(\lambda_k)| \right)^{1/j},$$

and the condition number reduction which is defined as

$$\psi \equiv \frac{\kappa(Q_{M^{-1}A})}{\kappa(Q_A)}, \tag{10.30}$$

where Q_A is the eigenvector matrix of A. From (10.3), we conclude that the smaller of ρ and ψ are, the faster the convergence rate will be. To determine the asymptotic convergence factor we have the following lemma by using Theorem 10.6.

Lemma 10.4 *Under the assumption in Theorem 10.6, all the eigenvalues given by (10.21) will asymptotically lie inside the interval*

$$\left[2, \quad 1 + 1/\sqrt{1 - \phi^2} \right].$$

Now, we prove the following lemma by using Lemma 10.4.

Lemma 10.5 *For $0 < \phi < \sqrt{8}/3$, the asymptotic convergence factor ρ_M satisfies*

$$\rho_M \leq \frac{1}{2}\left(-1 + \frac{1}{\sqrt{1 - \phi^2}}\right) < 1. \tag{10.31}$$

Proof: By Lemma 10.4, the eigenvalues given by (10.21) lie in a circle with the center $c = 2$ and the radius

$$R = -1 + \frac{1}{\sqrt{1 - \phi^2}}.$$

The asymptotic convergence factor ρ_M is bounded by R/c, see [75]. Hence,

$$\rho_M \leq \frac{R}{c} = \frac{1}{2}\left(-1 + \frac{1}{\sqrt{1 - \phi^2}}\right) < 1,$$

for $0 < \phi < \sqrt{8}/3$. $\square$

Under the condition in Lemma 10.5, it was proved in [44] that the semi-Toeplitz preconditioner T defined as in (10.29) gives rise to an asymptotic convergence factor

$$\rho_T \leq \frac{\sqrt{2 + 3\phi^2 - 2\sqrt{1 - \phi^2}}}{2}. \tag{10.32}$$

It can be verified that the bound in (10.31) is smaller than that in (10.32) when $0 < \phi < \sqrt{8}/3$. In Figure 10.3 we plot the bounds for ρ_M and ρ_T against the ϕ. We observe roughly that for $0 < \phi < 0.7$, $\rho_M \in [0, 0.2]$ while $\rho_T \in [0, 0.7]$. Therefore, it is expected that the preconditioner M gives a much faster convergence rate than that of T for most of the ϕ's. Let

$$\tilde{\rho} \equiv \left(\frac{\|r^{(i)}\|_2}{\|r^{(0)}\|_2} \right)^{1/i}$$

be the residual reduction. In Figure 10.4 we compare $\tilde{\rho}$ of the preconditioners M and T as a function of ϕ by using the GMRES method with $i = 10$. We observe that our preconditioner M indeed has a smaller residual reduction factor and hence leads to a faster convergence rate.

Next, we compare the condition number reduction ψ defined as in (10.30) of our preconditoner M with that of the semi-Toeplitz preconditioner T. Let us first compute the eigenvectors of $M^{-1}A$.

Theorem 10.8 *The preconditioned matrix $M^{-1}A$ has an invertible eigenvector matrix*

$$Q_{M^{-1}A} \equiv (V \otimes I_{m_1}) \text{diag}(U_{(1)}, \cdots, U_{(m_2)})$$

where V given by (10.9) is the eigenvector matrix of R_{m_2}, and

$$U_{(k)} \equiv \left(e_1, \cdots, e_{m_1-1}, \frac{u^{(k)}}{\|u^{(k)}\|_2} \right) \in \mathbb{C}^{m_1 \times m_1},$$

for $k = 1, \cdots, m_2$. Here, in $U_{(k)}$, $e_j \in \mathbb{R}^{m_1}$ is the j-th unit vector, for $j = 1, \cdots, m_1 - 1$, and the components of $u^{(k)} \in \mathbb{R}^{m_1}$ are given by

$$u_l^{(k)} = i^l(z_k^l - z_k^{-l}), \qquad i \equiv \sqrt{-1},$$

for $l = 1, \cdots, m_1$, where z_k is given by (10.20).

Proof: By (10.10), the matrix $M^{-1}E$ can be block diagonalized as follows,

$$(V \otimes I_{m_1})^{-1} M^{-1} E(V \otimes I_{m_1}) = \text{diag}(W_{(1)}, \cdots, W_{(m_2)}).$$

Owing to the structure of $W_{(k)}$, see (10.11), the last column of $W_{(k)}$, denoted by $u^{(k)}$, is the eigenvector corresponding to λ_k given by (10.13), and $e_1, \cdots, e_{m_1-1}$ are the eigenvectors corresponding to the zero eigenvalue of $W_{(k)}$. Therefore,

$$U_{(k)} \equiv \left(e_1, \cdots, e_{m_1-1}, \frac{u^{(k)}}{\|u^{(k)}\|_2} \right) \in \mathbb{C}^{m_1 \times m_1}$$

is the eigenvector matrix of $W_{(k)}$. We then obtain

$$U_{(k)}^{-1} W_{(k)} U_{(k)} = \Lambda_{(k)} = \text{diag}(0, \cdots, 0, \lambda_k),$$

for $k = 1, \cdots, m_2$. Moreover, the matrix

$$Q_{M^{-1}A} \equiv (V \otimes I_{m_1}) \text{diag}(U_{(1)}, \cdots, U_{(m_2)})$$

is invertible and

$$Q_{M^{-1}A}^{-1} M^{-1} A Q_{M^{-1}A} = I + Q_{M^{-1}A}^{-1} M^{-1} E Q_{M^{-1}A}$$

$$= I + \text{diag}(U_{(1)}^{-1} W_{(1)} U_{(1)}, \cdots, U_{(m_2)}^{-1} W_{(m_2)} U_{(m_2)})$$

$$= I + \text{diag}(\Lambda_{(1)}, \cdots, \Lambda_{(m_2)}),$$

i.e., $Q_{M^{-1}A}$ is the eigenvector matrix of $M^{-1}A$.

For the components of $u^{(k)}$, we have by Theorem 10.4, Lemma 10.3

and then Theorem 10.5,

$$u_l^{(k)} = \frac{2\kappa_1}{m_1} \sum_{j=1}^{m_1} \frac{i^l(-i)^{m_1}(-1)^{j+1}\sin\frac{(2j-1)l\pi}{2m_1}}{4 + \kappa_1\lambda_{1,j} + \kappa_2\lambda_{2,k}}$$

$$= \kappa_1 i^l(-i)^{m_1}\Phi_{m_1,l}\left(\frac{4 + \kappa_2\lambda_{2,k}}{2}, \kappa_1\right)$$

$$= \kappa_1 i^l(-i)^{m_1}\frac{i}{\kappa_1(z_k - z_k^{-1})}\left(z_k^{m_1+l} - z_k^{|m_1-l|}\right.$$

$$\left. - (z_k^{m_1} - z_k^{-m_1})(z_k^l - z_k^{-l})\frac{z_k^{2m_1}}{1 + z_k^{2m_1}}\right)$$

$$= \frac{i^{l+1}(-1)^{m_1}z_k^{m_1}}{z_k - z_k^{-1}}\left(1 - \frac{z_k^{m_1}}{1 + z_k^{2m_1}}(z_k^{m_1} - z_k^{-m_1})\right)(z_k^l - z_k^{-l}),$$

for $l = 1, \cdots, m_1$, where z_k is given by (10.20). By rescaling the column vector, we can choose $u^{(k)}$, for $k = 1, \cdots, m_2$, as the eigenvectors with the components given by

$$u_l^{(k)} = i^l(z_k^l - z_k^{-l}),$$

for $l = 1, \cdots, m_1$. $\square$

It was shown in [44] that the condition number reduction ψ defined by (10.30) satisfies

$$\psi \le \psi_0 \equiv \max_{1 \le k \le m_2}\|U_{(k)}\|_2 \cdot \max_{1 \le k \le m_2}\|U_{(k)}^{-1}\|_2/\kappa(H)$$

where H is the eigenvector matrix of $4I_{m_1} + \kappa_1 R_{m_1}$ and R_{m_1} is given by (10.7). Therefore, we only need to determine $\|U_{(k)}\|_2$ and $\|U_{(k)}^{-1}\|_2$ in order to estimate ψ_0.

Theorem 10.9 *We have*

$$\|U_{(k)}\|_2 = \left(1 + \sqrt{1 - a_k}\right)^{1/2}$$

and

$$\|U_{(k)}^{-1}\|_2 = \left(1 - \sqrt{1 - a_k}\right)^{-1/2},$$

for $k = 1, \cdots, m_2$, where

$$a_k \equiv \frac{|z_k^{m_1} - z_k^{-m_1}|^2}{\dfrac{|z_k|^2 - |z_k|^{2m_1+2}}{1 - |z_k|^2} + \dfrac{|z_k|^{-2} - |z_k|^{-2m_1-2}}{1 - |z_k|^{-2}} - 2Re\left(\dfrac{z_k \bar{z}_k^{-1} - (z_k \bar{z}_k^{-1})^{m_1+1}}{1 - z_k \bar{z}_k^{-1}}\right)}$$

(10.33)

with z_k defined as in (10.20).

Proof: Since

$$U_{(k)}^* U_{(k)} = \begin{pmatrix} I_{m_1-1} & v^{(k)} \\ v^{(k)*} & 1 \end{pmatrix}$$

where $v^{(k)} \in \mathbb{R}^{m_1-1}$ with components given by

$$v_l^{(k)} = \frac{1}{\|u^{(k)}\|_2} u_l^{(k)},$$

for $l = 1, \cdots, m_1 - 1$, we have

$$U_{(k)}^* U_{(k)} - \lambda I = \begin{pmatrix} (1-\lambda)I_{m_1-1} & v^{(k)} \\ v^{(k)*} & 1 - \lambda \end{pmatrix}$$

$$= \begin{pmatrix} (1-\lambda)I_{m_1-1} & 0 \\ v^{(k)*} & (1-\lambda) - \frac{1}{1-\lambda} v^{(k)*} v^{(k)} \end{pmatrix}$$

$$\times \begin{pmatrix} I_{m_1-1} & \frac{1}{1-\lambda} v^{(k)} \\ 0 & 1 \end{pmatrix}.$$

The characteristic polynomial of $U_{(k)}^* U_{(k)}$ is given by

$$(1-\lambda)^{m_1-2}[(1-\lambda)^2 - \|v^{(k)}\|_2^2] = 0.$$

Therefore, the eigenvalues of $U_{(k)}^* U_{(k)}$ are 1 with multiplicity $m_1 - 2$ and the other two are

$$1 + \|v^{(k)}\|_2, \qquad 1 - \|v^{(k)}\|_2.$$

Hence,

$$\|U_{(k)}\|_2 = (1 + \|v^{(k)}\|_2)^{1/2}$$

and

$$\|U_{(k)}^{-1}\|_2 = (1 - \|v^{(k)}\|_2)^{-1/2},$$

for $k = 1, \cdots, m_2$. For $\|v^{(k)}\|_2^2$, we have

$$\|v^{(k)}\|_2^2 = \sum_{l=1}^{m_1-1} |v_l^{(k)}|^2 = 1 - \frac{|u_{m_1}^{(k)}|^2}{\|u^{(k)}\|_2^2} = 1 - \frac{|z_k^{m_1} - z_k^{-m_1}|^2}{\|u^{(k)}\|_2^2}$$

$$= 1 - \frac{|z_k^{m_1} - z_k^{-m_1}|^2}{\frac{|z_k|^2 - |z_k|^{2m_1+2}}{1 - |z_k|^2} + \frac{|z_k|^{-2} - |z_k|^{-2m_1-2}}{1 - |z_k|^{-2}} - 2Re\left(\frac{z_k \bar{z}_k^{-1} - (z_k \bar{z}_k^{-1})^{m_1+1}}{1 - z_k \bar{z}_k^{-1}}\right)}$$

$$= 1 - a_k$$

where a_k is given in (10.33). □

In Figure 10.5 we compare the upper bound ψ_0 of the condition number reduction of our preconditoner M with that of the semi-Toeplitz preconditioner T. We see that ψ_0 of both M and T decreases when the size of system increases.

Finally, Figures 10.6 and 10.7 give the number of iterations obtained from the restarted GMRES(10) method (see [75]) for different mesh sizes. The iteration is stopped when

$$\frac{\|r^{(k)}\|_2}{\|r^{(0)}\|_2} < 10^{-6}$$

where $r^{(k)}$ is the residual vector at the k-th iteration. We observe that M as a preconditioner is more effective than T for various ϕ and α. Also, the number of iterations for convergence decreases when ϕ decreases which agrees with Lemma 10.4 and Figure 10.3.

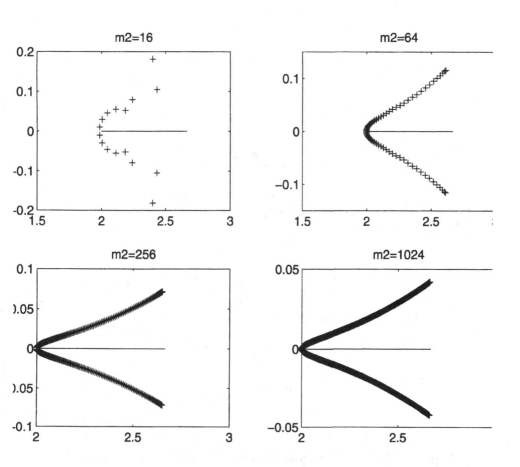

Figure 10.1   +: eigenvalues (10.21),   −: asymptotic spectrum,

$$\phi = 0.8, \quad c = 10, \quad \alpha = 0.6.$$

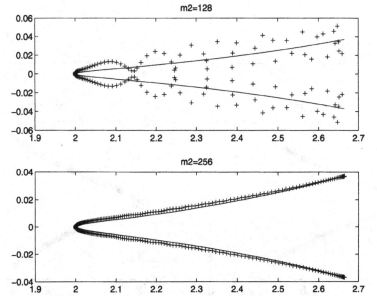

Figure 10.2   +: eigenvalues (10.21),   −: asymptotic
spectrum,

$$\phi = 0.8, \quad \kappa_1 = 200.$$

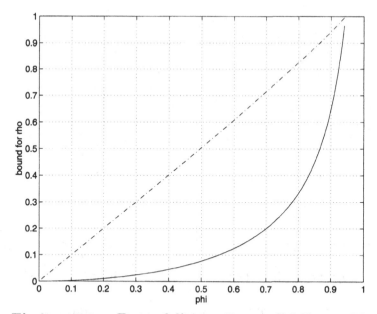

Figure 10.3   **Dotted line $= T$,   solid line $= M$.**

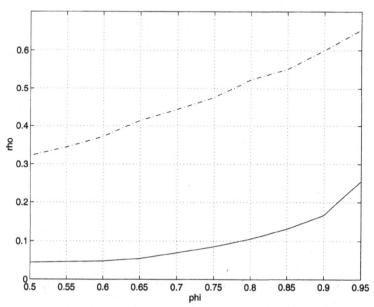

Figure 10.4   **Dotted line $= T$,   solid line $= M$,**

$$m_1 = 128, \quad c = 100, \quad \alpha = 0.99.$$

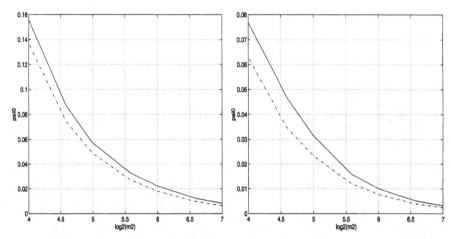

Figure 10.5   Left: $\phi = 0.99$,   **right:** $\phi = 0.5$,

dotted line $= T$, solid line $= M$, $\alpha = 0.99$, $c = 100$.

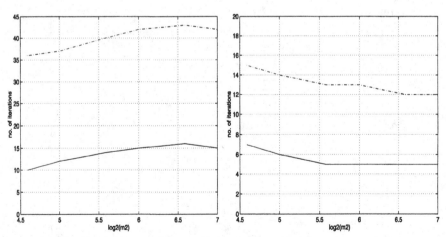

Figure 10.6   Left: $\phi = 0.99$,   **right:** $\phi = 0.5$,

dotted line $= T$, solid line $= M$, $\alpha = 0.99$, $c = 100$.

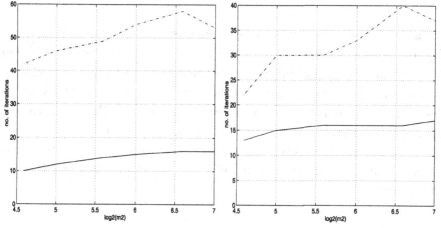

Figure 10.7   **Left:** $\phi = 0.99$,   **right:** $\phi = 0.5$,

dotted line $= T$, solid line $= M$, $\alpha = 0.1$, $c = 10$.

Chapter 11

Applications in ODEs and DAEs

We study the solutions of linear systems arising from numerical methods for ordinary differential equations (ODEs). Since the matrix of the system is non-symmetric, large and sparse, we then use the GMRES method with a block circulant preconditioner. The eigenvalues of the preconditioned system are clustered around 1. It follows that when the GMRES method is applied to solving the preconditioned system, we expect a fast convergence rate. Numerical results are given to illustrate the effectiveness of our method. An algorithm for solving differential algebraic equations (DAEs) is also given

11.1 BVMs and their matrix forms

We consider the linear initial value problem (IVP)

$$
\begin{cases}
\dfrac{dy(t)}{dt} = J_m y(t) + g(t), \quad t \in (t_0, T], \\[2mm]
y(t_0) = \eta,
\end{cases}
\tag{11.1}
$$

where $y(t), g(t) : \mathbb{R} \to \mathbb{R}^m$, $\eta \in \mathbb{R}^m$ and $J_m \in \mathbb{R}^{m \times m}$. For solving (11.1), usually, we use initial value methods (IVMs). Recently, a class of boundary value methods (BVMs) has been introduced, see [13]. Such methods are based on linear multistep formulae (LMF). For the linear IVP (11.1), a

BVM approximates the solution of (11.1) by means of a discrete boundary value problem (BVP). By using a k-step LMF over uniform grid points

$$t_j = t_0 + jh, \quad j = 0, \cdots, s,$$

with $h = \dfrac{T - t_0}{s}$, we have a discrete BVP as follows,

$$\sum_{i=-\nu}^{k-\nu} \alpha_{i+\nu} y_{n+i} = h \sum_{i=-\nu}^{k-\nu} \beta_{i+\nu} f_{n+i}, \quad n = \nu, \ldots, s - k + \nu. \qquad (11.2)$$

Here, as usual, y_n is the discrete approximation to $y(t_n)$,

$$f_n = J_m y_n + g_n$$

with $g_n = g(t_n)$. We note that the initial condition in (11.1) only gives one value

$$y_0 = y(t_0) = \eta$$

to us. Therefore, in order to solve the BVP defined as in (11.2), we need the values

$$y_1, y_2, \cdots, y_{\nu-1},$$

and

$$y_{s-k+\nu+1}, \cdots, y_s.$$

With additional conditions which provide the following set of $k - 1$ equations (for simplicity, such formulae are assumed to have the same number of steps as the main method), we have

$$\sum_{i=0}^{k} \alpha_i^{(j)} y_i = h \sum_{i=0}^{k} \beta_i^{(j)} f_i, \quad j = 1, \ldots, \nu - 1, \qquad (11.3)$$

and

$$\sum_{i=0}^{k} \alpha_{k-i}^{(j)} y_{s-i} = h \sum_{i=0}^{k} \beta_{k-i}^{(j)} f_{s-i}, \quad j = s - k + \nu + 1, \ldots, s. \qquad (11.4)$$

The equations in (11.3) are called the additional initial conditions and the equations in (11.4) are called the additional final conditions. We then have a well-posed linear system with the equations (11.2), (11.3) and

(11.4). This linear system determines the use of a BVM on the problem (11.1).

The BVM considered here is assumed to be consistent, i.e., it satisfies the conditions

$$\rho(1) = 0 \quad \text{and} \quad \rho'(1) = \sigma(1)$$

where $\rho(z)$ and $\sigma(z)$ denote two characteristic polynomials associated with the given method, i.e.,

$$\rho(z) \equiv z^\nu \sum_{j=-\nu}^{k-\nu} \alpha_{j+\nu} z^j \tag{11.5}$$

and

$$\sigma(z) \equiv z^\nu \sum_{j=-\nu}^{k-\nu} \beta_{j+\nu} z^j. \tag{11.6}$$

Now, we compare BVMs with IVMs briefly. For traditional IVMs, we know that the system of equations can be solved easily by forward recursions. Although IVMs are more efficient than BVMs, the advantage in using BVMs over IVMs is that the BVMs have much better stability properties which are important in the numerical process, see [13].

The BVMs in matrix form are given by introducing $\widehat{A}$, $\widehat{B} \in \mathbb{R}^{(s+1)\times(s+1)}$ as follows,

$$\widehat{A} \equiv \begin{pmatrix} 1 & \cdots & 0 & & & & & \\ \alpha_0^{(1)} & \cdots & \alpha_k^{(1)} & & & & & \\ \vdots & \vdots & \vdots & & & 0 & & \\ \alpha_0^{(\nu-1)} & \cdots & \alpha_k^{(\nu-1)} & & & & & \\ \alpha_0 & \cdots & \alpha_k & & & & & \\ & \alpha_0 & \cdots & & \alpha_k & & & \\ & & \ddots & & \ddots & \ddots & & \\ & & & \ddots & & \ddots & & \ddots \\ & & & \alpha_0 & \cdots & \alpha_k & & \\ & 0 & & \alpha_0^{(s-k+\nu+1)} & \cdots & \alpha_k^{(s-k+\nu+1)} & & \\ & & & \vdots & & \vdots & \vdots & \\ & & & \alpha_0^{(s)} & \cdots & \alpha_k^{(s)} & & \end{pmatrix} \tag{11.7}$$

and

$$\widehat{B} \equiv \begin{pmatrix} 0 & \cdots & 0 & & & & & & \\ \beta_0^{(1)} & \cdots & \beta_k^{(1)} & & & & & \\ \vdots & \vdots & \vdots & & & & 0 & \\ \beta_0^{(\nu-1)} & \cdots & \beta_k^{(\nu-1)} & & & & & \\ \beta_0 & \cdots & \beta_k & & & & & \\ & \beta_0 & \cdots & & \beta_k & & & \\ & & \ddots & & \ddots & \ddots & & \\ & & & \ddots & \ddots & & \ddots & \\ & & & & \beta_0 & \cdots & \beta_k & \\ & 0 & & & \beta_0^{(s-k+\nu+1)} & \cdots & \beta_k^{(s-k+\nu+1)} \\ & & & & \vdots & \vdots & \vdots \\ & & & & \beta_0^{(s)} & \cdots & \beta_k^{(s)} \end{pmatrix}. \tag{11.8}$$

Thus, the discrete problem generated by the application of the BVM (11.2)–(11.4) to the problem (11.1) is given by

$$\widehat{M}\boldsymbol{y} = e_1 \otimes \eta + h(\widehat{B} \otimes I_m)\boldsymbol{g} \tag{11.9}$$

where

$$\widehat{M} \equiv \widehat{A} \otimes I_m - h\widehat{B} \otimes J_m \in \mathbb{R}^{(s+1)m \times (s+1)m}, \tag{11.10}$$

$$e_1 = (1, 0, \cdots, 0)^T \in \mathbb{R}^{(s+1)},$$

$$\boldsymbol{y} = (y_0, \cdots, y_s)^T \in \mathbb{R}^{(s+1)m}$$

and

$$\boldsymbol{g} = (g_0, \cdots, g_s)^T \in \mathbb{R}^{(s+1)m}.$$

For simplicity, J_m here is assumed to be invertible. For instance, J_m will be the discrete Laplacian matrix for heat equations and the Hamiltonian matrix for wave equations, see numerical examples in Section 11.4. The matrix $\widehat{M}$ in (11.10) turns out to be large and sparse when either $s \gg k$ or J_m is large and sparse. In those cases the solution of (11.9) by using a direct method is expensive. We therefore consider the use of iterative methods, such as the GMRES method, see Section 10.1 or [38, 75], with a suitable preconditioner.

11.2 Construction of preconditioner

We note that the matrices $\widehat{A}$ given by (11.7) and $\widehat{B}$ given by (11.8) are Toeplitz matrices with low rank perturbations. By neglecting the low rank perturbations, we have the following band-Toeplitz matrices,

$$
A \equiv
\begin{pmatrix}
\alpha_\nu & \cdots & \alpha_k & 0 & \cdots & 0 \\
\vdots & \alpha_\nu & \cdots & \alpha_k & 0 & \\
\alpha_0 & \vdots & \ddots & & \ddots & \ddots \\
0 & \alpha_0 & & \ddots & & \alpha_k \\
\vdots & 0 & \ddots & & \ddots & \vdots \\
0 & & \ddots & \alpha_0 & \cdots & \alpha_\nu
\end{pmatrix}
\in \mathbb{R}^{(s+1)\times(s+1)}
$$

and

$$
B \equiv
\begin{pmatrix}
\beta_\nu & \cdots & \beta_k & 0 & \cdots & 0 \\
\vdots & \beta_\nu & \cdots & \beta_k & 0 & \\
\beta_0 & \vdots & \ddots & & \ddots & \ddots \\
0 & \beta_0 & & \ddots & & \beta_k \\
\vdots & 0 & \ddots & & \ddots & \vdots \\
0 & & \ddots & \beta_0 & \cdots & \beta_\nu
\end{pmatrix}
\in \mathbb{R}^{(s+1)\times(s+1)}.
$$

The matrix

$$
M \equiv A \otimes I_m - hB \otimes J_m
$$

is a block Toeplitz approximation to the matrix $\widehat{M}$. We then construct the following block circulant preconditioner for M (and hence $\widehat{M}$),

$$
\tilde{s}_F^{(1)}(M) = s(A) \otimes I_m - hs(B) \otimes J_m. \tag{11.11}
$$

Here $s(A)$, $s(B) \in \mathbb{R}^{(s+1)\times(s+1)}$ are circulant matrices given by

$$s(A) = \begin{pmatrix} \alpha_\nu & \cdots & \alpha_k & 0 & \cdots\cdots & 0 & \alpha_0 & \cdots & \alpha_{\nu-1} \\ \vdots & & \alpha_\nu & \cdots & \alpha_k & 0 & & \ddots & \vdots \\ \alpha_0 & & \vdots & \ddots & & \ddots & \ddots & & \alpha_0 \\ 0 & & \alpha_0 & & \ddots & & \ddots & \ddots & 0 \\ \vdots & & 0 & \ddots & & \ddots & & & \vdots \\ 0 & & & & \ddots & \ddots & & & 0 \\ \alpha_k & & & \ddots & & \ddots & & & \alpha_k \\ \vdots & & \ddots & & & & & & \vdots \\ \alpha_{\nu+1} & \cdots & \alpha_k & 0 & \cdots\cdots & 0 & \alpha_0 & \cdots & \alpha_\nu \end{pmatrix}$$

and

$$s(B) = \begin{pmatrix} \beta_\nu & \cdots & \beta_k & 0 & \cdots\cdots & 0 & \beta_0 & \cdots & \beta_{\nu-1} \\ \vdots & & \beta_\nu & \cdots & \beta_k & 0 & & \ddots & \vdots \\ \beta_0 & & \vdots & \ddots & & \ddots & \ddots & & \beta_0 \\ 0 & & \beta_0 & & \ddots & & \ddots & \ddots & 0 \\ \vdots & & 0 & \ddots & & \ddots & & & \vdots \\ 0 & & & & \ddots & \ddots & & & 0 \\ \beta_k & & & \ddots & & \ddots & & & \beta_k \\ \vdots & & \ddots & & & & & & \vdots \\ \beta_{\nu+1} & \cdots & \beta_k & 0 & \cdots\cdots & 0 & \beta_0 & \cdots & \beta_\nu \end{pmatrix}.$$

Recall that $\{\alpha_i\}_{i=0}^k$ and $\{\beta_i\}_{i=0}^k$ are the coefficients in the LMF, see (11.2). We note that $s(A)$ and $s(B)$ are just Strang's circulant preconditioners for A and B respectively, see Section 1.2.1. The preconditioner $\tilde{s}_F^{(1)}(M)$ is Strang's block circulant preconditioner for M proposed in Section 2.3.2. The stability of a BVM is closely related to the stability polynomial defined by

$$\pi(z,q) \equiv \rho(z) - q\sigma(z), \quad q \in \mathbb{C},$$

where $\rho(z)$ and $\sigma(z)$ are given by (11.5) and (11.6) respectively. Therefore we have the following definition, see [13].

Definition 11.1 *The A_{k_1,k_2}-stable region $\mathcal{D}_{k_1,k_2}$ in $\mathbb{C}$ is defined by*

$$\mathcal{D}_{k_1,k_2} = \{q \in \mathbb{C} \mid \pi(z,q) \text{ has } k_1 \text{ zeros inside } |z| = 1$$

and k_2 zeros outside $|z| = 1$}.

A k-step BVM with k_1 initial conditions and k_2 final conditions ($k = k_1 + k_2$) is said to be A_{k_1,k_2}-stable if

$$\mathbb{C}^- \subseteq \mathcal{D}_{k_1,k_2}$$

where $\mathbb{C}^- = \{q \in \mathbb{C}^\cdot | \ Re(q) < 0\}$.

By using a A_{k_1,k_2}-stable BVM for solving ODEs we have the following invertibility result for our preconditioner $\tilde{s}_F^{(1)}(M)$ with $k_1 = \nu$ and $k_2 = k - \nu$.

Theorem 11.1 *If a BVM for the IVP (11.1) is $A_{\nu,k-\nu}$-stable and $h\mu_k \in \mathcal{D}_{\nu,k-\nu}$ where μ_k, for $k = 1, \cdots, m$, are the eigenvalues of J_m, then the preconditioner*

$$\tilde{s}_F^{(1)}(M) = s(A) \otimes I_m - hs(B) \otimes J_m$$

is invertible.

Proof: We know that the eigenvalues of $s(A)$ are given by $p_A(\omega_j)$ with $\omega_j = e^{\frac{2\pi ij}{s+1}}$, for $j = 0, \cdots, s$, where

$$p_A(z) = \alpha_k z^{k-\nu} + \cdots + \alpha_\nu + \alpha_{\nu-1}\frac{1}{z} + \cdots + \alpha_0\frac{1}{z^\nu}.$$

The eigenvalues of $s(B)$ are given by $p_B(\omega_j)$, for $j = 0, \cdots, s$, where

$$p_B(z) = \beta_k z^{k-\nu} + \cdots + \beta_\nu + \beta_{\nu-1}\frac{1}{z} + \cdots + \beta_0\frac{1}{z^\nu},$$

see [33] and [40]. Then, the eigenvalues of

$$s(A) \otimes I_m - hs(B) \otimes J_m$$

are

$$\lambda_{j,k} = p_A(\omega_j) - h\mu_k p_B(\omega_j), \tag{11.12}$$

for $j = 0, \cdots, s$, and $k = 1, \cdots, m$, where μ_k are the eigenvalues of J_m. Since

$$p_A(z) + qp_B(z) = \frac{1}{z^\nu}(\rho(z) + q\sigma(z)) = \frac{1}{z^\nu}\pi(z,q), \tag{11.13}$$

we have by the assumption of $A_{\nu,k-\nu}$-stability,

$$\pi(\omega, q) = \rho(\omega) + q\sigma(\omega) \neq 0, \quad \forall |\omega| = 1, \quad q \in \mathcal{D}_{\nu,k-\nu}. \qquad (11.14)$$

If $h\mu_k \in \mathcal{D}_{\nu,k-\nu}$, then by (11.12), (11.13) and (11.14), we have

$$\lambda_{j,k} \neq 0,$$

for $j = 0, \cdots, s$, and $k = 1, \cdots, m$. Thus the matrix $\tilde{s}_F^{(1)}(M)$ is invertible.
□

We remark that if a BVM is $A_{\nu,k-\nu}$-stable and $\mu_k \in \mathbb{C}^-$, then $\tilde{s}_F^{(1)}(M)$ is invertible by noting that $h\mu_k \in \mathbb{C}^- \subseteq \mathcal{D}_{\nu,k-\nu}$.

11.3 Convergence rate and operation cost

In this section we want to show that the spectrum of the preconditioned system is clustered.

Theorem 11.2 *All the eigenvalues of the preconditioned matrix* $(\tilde{s}_F^{(1)}(M))^{-1}\widehat{M}$ *are 1 except for at most 2mk outliers.*

Proof: Let

$$E \equiv \widehat{M} - \tilde{s}_F^{(1)}(M).$$

We have by (11.10) and (11.11),

$$E = (\widehat{A} - s(A)) \otimes I_m - h(\widehat{B} - s(B)) \otimes J_m = L_A \otimes I_m - hL_B \otimes J_m. \quad (11.15)$$

For L_A, $L_B \in \mathbb{R}^{(s+1)\times(s+1)}$ in (11.15), we know that they have non-zero entries only in the following four corners:

(i) $\nu \times (k+1)$ in the upper left;

(ii) $\nu \times \nu$ upper right;

(iii) $(k - \nu) \times (k+1)$ lower right;

(iv) $(k - \nu) \times (k - \nu)$ lower left.

Note that $k > \nu$, $\text{rank}(L_A) \leq k$, and $\text{rank}(L_B) \leq k$. We then have

$$\text{rank}(L_A \otimes I_m) = \text{rank}(L_A) \cdot m \leq mk,$$

and

$$\text{rank}(L_B \otimes J_m) = \text{rank}(L_B) \cdot m \leq mk.$$

Since $\tilde{s}_F^{(1)}(M)$ is invertible we therefore have

$$(\tilde{s}_F^{(1)}(M))^{-1}\widehat{M} = I + (s_F^{(1)}(M))^{-1}E = I + L \qquad (11.16)$$

with $\text{rank}(L) \leq 2mk$. $\square$

If the GMRES method is applied to solving the preconditioned system, we then have the following corollary by using (11.16) and Lemma 10.1.

Corollary 11.1 *When the GMRES method is applied to solving the preconditioned system, the method will converge in at most $2mk+1$ iterations in exact arithmetic.*

Finally, we consider the computational cost of our method. The number of operations per iteration in the preconditioned GMRES method depends mainly on the work of computing the matrix-vector multiplication

$$(\tilde{s}_F^{(1)}(M))^{-1}\widehat{M}\mathbf{z} = (s(A) \otimes I_m - hs(B) \otimes J_m)^{-1}(\widehat{A} \otimes I_m - h\widehat{B} \otimes J_m)\mathbf{z},$$

see for instance, Saad [75]. Since $\widehat{A}$, $\widehat{B}$ are band matrices and J_m is assumed to be sparse, the matrix-vector multiplication

$$\widehat{M}\mathbf{z} = (\widehat{A} \otimes I_m - h\widehat{B} \otimes J_m)\mathbf{z}$$

can be done very fast.

To compute $(\tilde{s}_F^{(1)}(M))^{-1}(\widehat{M}\mathbf{z})$, since $s(A)$ and $s(B)$ are circulant matrices, we have

$$s(A) = F^*\Lambda_A F \quad \text{and} \quad s(B) = F^*\Lambda_B F$$

where Λ_A, Λ_B are diagonal matrices and F is the Fourier matrix, see (1.6). It follows that

$$(\tilde{s}_F^{(1)}(M))^{-1}(\widehat{M}\mathbf{z}) = (F^* \otimes I_m)(\Lambda_A \otimes I_m - h\Lambda_B \otimes J_m)^{-1}(F \otimes I_m)(\widehat{M}\mathbf{z}).$$

This product can be obtained by using FFTs and solving s linear systems of order m. Since J_m is sparse, the coefficient matrices of the $m \times m$ linear systems will also be sparse. Thus,

$$(\tilde{s}_F^{(1)}(M))^{-1}(\widehat{M}\mathbf{z})$$

can be obtained by solving s sparse $m \times m$ linear systems.

11.4 Algorithm for DAEs

Let us consider the linear differential algebraic equation (DAE)

$$\begin{cases} A\dfrac{dx(t)}{dt} + Bx(t) = f(t), \quad t \in (0,1], \\[2mm] x(0) = \eta, \end{cases} \tag{11.17}$$

where $A, B \in \mathbb{R}^{m \times m}$, A is not invertible, $x(t) : \mathbb{R} \to \mathbb{R}^m$ and $\eta \in \mathbb{R}^m$. Such DAEs arise in a variety of applications in electrical engineering and control theory, see [11].

A matrix pencil is defined by $\lambda A + B$ with $\lambda \in \mathbb{C}$. The pencil is said to be regular if

$$\det(\lambda A + B) \not\equiv 0$$

as a function of λ. If the pencil $\lambda A + B$ is regular, then the equation (11.17) is solvable and there exist two invertible matrices P and Q such that

$$PAQ = \begin{pmatrix} I & 0 \\ 0 & N \end{pmatrix}, \qquad PBQ = \begin{pmatrix} G & 0 \\ 0 & I \end{pmatrix},$$

where N is nilpotent of order ν, i.e., $N^\nu = 0$ and $N^{\nu-1} \neq 0$, see [11]. The order of the nilpotency ν is called the index of the matrix pencil. We remark that the matrices P and Q are not easy to be found in general. In [84], there is a constructive approach to calculate P and Q. The scheme is given as follows.

(i) Let $B_{(1)} = cA + B$ be invertible for some $c \in \mathbb{C}$. Then for any $\lambda \in \mathbb{C}$, we have

$$B_{(1)}^{-1}(\lambda A + B) = B_{(1)}^{-1}(B_{(1)} + (\lambda - c)A) = I + (\lambda - c)B_{(1)}^{-1}A.$$

(ii) Let N be an invertible matrix such that $N^{-1}B_{(1)}^{-1}AN$ is of the Jordan canonical form, see Theorem 1.5 in Section 1.1.1. By interchanging the columns of N we can assume that

$$N^{-1}B_{(1)}^{-1}AN = \text{diag}(J_{(1)}, J_{(0)}),$$

where $J_{(1)}$ and $J_{(0)}$ are Jordan blocks. All the entries in the main diagonal of $J_{(1)}$ are non-zero and all the entries in the main diagonal of $J_{(0)}$ are zeros. Then we have

$$N^{-1}(I + (\lambda - c)B_{(1)}^{-1}A)N = \begin{pmatrix} I + (\lambda - c)J_{(1)} & 0 \\ 0 & (I - cJ_{(0)}) + \lambda J_{(0)} \end{pmatrix}.$$

(iii) Compute

$$\begin{pmatrix} I & 0 \\ 0 & (I - cJ_{(0)})^{-1} \end{pmatrix} \begin{pmatrix} I + (\lambda - c)J_{(1)} & 0 \\ 0 & (I - cJ_{(0)}) + \lambda J_{(0)} \end{pmatrix}$$
$$= \begin{pmatrix} I + (\lambda - c)J_{(1)} & 0 \\ 0 & I + \lambda(I - cJ_{(0)})^{-1}J_{(0)} \end{pmatrix}.$$

(iv) Since $J_{(0)}$ is nilpotent and $(I - cJ_{(0)})^{-1}$ commutes with $J_{(0)}$, the matrix

$$(I - cJ_{(0)})^{-1}J_{(0)}$$

is also nilpotent. Let S be an invertible matrix such that

$$S^{-1}(I - cJ_{(0)})^{-1}J_{(0)}S = N$$

is of the Jordan canonical form. Then we have

$$\begin{pmatrix} I & 0 \\ 0 & S^{-1} \end{pmatrix} \begin{pmatrix} I + (\lambda - c)J_{(1)} & 0 \\ 0 & I + \lambda(I - cJ_{(0)})^{-1}J_{(0)} \end{pmatrix} \begin{pmatrix} I & 0 \\ 0 & S \end{pmatrix}$$
$$= \begin{pmatrix} I + (\lambda - c)J_{(1)} & 0 \\ 0 & I + \lambda N \end{pmatrix}.$$

(v) Compute

$$
\begin{pmatrix} J_{(1)}^{-1} & 0 \\ 0 & I \end{pmatrix} \begin{pmatrix} I + (\lambda - c)J_{(1)} & 0 \\ 0 & I + \lambda N \end{pmatrix}
$$

$$
= \begin{pmatrix} J_{(1)}^{-1} + (\lambda - c)I & 0 \\ 0 & I + \lambda N \end{pmatrix}
$$

$$
= \begin{pmatrix} G & 0 \\ 0 & I \end{pmatrix} + \lambda \begin{pmatrix} I & 0 \\ 0 & N \end{pmatrix}
$$

where $G = J_{(1)}^{-1} - cI$.

(vi) Let P be the product of all the matrices used to multiply the matrix pencil $\lambda A + B$ on the left in (i)–(v) and let Q be the product of all the matrices used to multiply the matrix pencil on the right in (i)–(v). Then

$$
P(\lambda A + B)Q = \lambda \begin{pmatrix} I & 0 \\ 0 & N \end{pmatrix} + \begin{pmatrix} G & 0 \\ 0 & I \end{pmatrix}.
$$

These P and Q are our desired matrices.

Applying the coordinate changes P and Q to the DAE defined as in (11.17), we have

$$
\begin{cases} \dfrac{dy_1(t)}{dt} + Gy_1(t) = g_1(t), \\[3mm] N\dfrac{dy_2(t)}{dt} + y_2(t) = g_2(t), \end{cases}
$$

where

$$
y = \begin{pmatrix} y_1 \\ y_2 \end{pmatrix} = Q^{-1}x \quad \text{and} \quad g = \begin{pmatrix} g_1 \\ g_2 \end{pmatrix} = Pf.
$$

We note that

$$
\frac{dy_1(t)}{dt} + Gy_1(t) = g_1(t)
$$

is a system of ODEs defined as in (11.1) and a solution exists for any initial value of $y_1(t)$. Therefore, we can use the method proposed in previous sections to solve this system of ODEs. For

$$
N\frac{dy_2(t)}{dt} + y_2(t) = g_2(t),
$$

we know that there exists only one solution

$$y_2(t) = \sum_{k=0}^{\nu-1} (-1)^k N^k g_2^{(k)}(t)$$

where $g_2^{(k)}(t) = \dfrac{d^k g_2(t)}{dt^k}$.

11.5   Numerical results

By numerical tests for solving ODE and DAE problems, we illustrate the efficiency of employing the preconditioner $\tilde{s}_F^{(1)}(M)$ over other circulant type of preconditioners. The GMRES method, see Section 10.1, is applied to solving preconditioned systems arising from numerical implicit methods for ODEs. In our tests the zero vector is the initial guess and the stopping criterion is

$$\frac{||r^{(k)}||_2}{||r^{(0)}||_2} < 10^{-6}$$

where $r^{(k)}$ is the residual vector after k-th iterations.

Example 1. Heat equation:

$$
\begin{cases}
\dfrac{\partial u}{\partial t} - \dfrac{\partial^2 u}{\partial x^2} = 0, \\[2mm]
u(0,t) = u(\pi,t) = 0, \quad t \in [0, 2\pi], \\[2mm]
u(x,0) = \sin x, \qquad x \in [0,\pi].
\end{cases}
\tag{11.18}
$$

We discretize the operator $\partial^2/\partial x^2$ in (11.18) by using the centered difference with a step $\pi/(m+1)$ and therefore obtain a system of ODEs,

$$
\begin{cases}
\dfrac{dy(t)}{dt} = T_m y(t), \qquad\qquad t \in [0, 2\pi], \\[2mm]
y(0) = (\sin x_1, \sin x_2, \cdots, \sin x_m)^T,
\end{cases}
\tag{11.19}
$$

where

$$
T_m = \frac{(m+1)^2}{\pi^2}
\begin{pmatrix}
-2 & 1 & & \\
1 & \ddots & \ddots & \\
& \ddots & \ddots & 1 \\
& & 1 & -2
\end{pmatrix}
\in \mathbb{R}^{m \times m}.
\tag{11.20}
$$

The third order generalized backward differentiation formula (GBDF) is used to solve (11.19). The formulae and the additional initial and final conditions can be found in [13].

Example 2. Wave equation:

$$
\begin{cases}
\dfrac{\partial^2 u}{\partial t^2} - \dfrac{\partial^2 u}{\partial x^2} = 0, \\[2mm]
u(0,t) = u(\pi, t) = 0, & t \in [0, 2\pi], \\[2mm]
u(x, 0) = x, & x \in [0, \pi], \\[2mm]
\dfrac{\partial u}{\partial t}(x, 0) = 0, & x \in [0, \pi].
\end{cases}
\tag{11.21}
$$

We again discretize the operator $\partial^2/\partial x^2$ in (11.21) by using the centered difference with a step $\pi/(m+1)$ and therefore obtain a system of ODEs,

$$
\begin{cases}
\dfrac{dy(t)}{dt} = H_m y(t), & t \in [0, 2\pi], \\[2mm]
y(0) = (x_1, x_2, \cdots, x_{m/2}, 0, \cdots, 0)^T,
\end{cases}
\tag{11.22}
$$

where

$$
H_m = \begin{pmatrix} 0_{m/2} & I_{m/2} \\ T_{m/2} & 0_{m/2} \end{pmatrix} \in \mathbb{R}^{m \times m}
$$

(assume m is even) with $T_{m/2}$ given by (11.20). The fourth order extended trapezoidal rule of second kind (ETR$_2$) is used to solve (11.22). The formulae and the additional initial and final conditions can also be found in [13].

Example 3. DAE:

$$
\begin{cases}
A\dfrac{dx(t)}{dt} + Bx(t) = 0, & t \in (0, 1], \\[2mm]
x(0) = (1, \cdots, 1),
\end{cases}
$$

where

$$A = \begin{pmatrix} 50 & 114 & 95 & 140 & 129 & 91 & 43 \\ 101 & 198 & 149 & 155 & 223 & 183 & 138 \\ 97 & 206 & 156 & 197 & 187 & 156 & 87 \\ 82 & 185 & 148 & 164 & 156 & 129 & 81 \\ 82 & 202 & 167 & 186 & 201 & 180 & 114 \\ 111 & 226 & 193 & 197 & 229 & 198 & 138 \\ 32 & 122 & 107 & 100 & 115 & 100 & 74 \end{pmatrix}$$

and

$$B = \begin{pmatrix} 118 & 229 & 242 & 318 & 278 & 94 & 61 \\ 193 & 470 & 333 & 344 & 457 & 325 & 266 \\ 286 & 458 & 379 & 470 & 379 & 156 & 102 \\ 304 & 474 & 375 & 429 & 429 & 240 & 181 \\ 210 & 459 & 407 & 411 & 432 & 295 & 230 \\ 306 & 588 & 484 & 512 & 529 & 385 & 277 \\ 100 & 295 & 257 & 197 & 338 & 265 & 246 \end{pmatrix} .$$

There exist two invertible matrices

$$P^{-1} = \begin{pmatrix} 9 & 0 & 1 & 3 & 2 & 4 & 6 \\ 2 & 8 & 4 & 8 & 1 & 8 & 3 \\ 6 & 4 & 9 & 0 & 0 & 5 & 8 \\ 4 & 6 & 9 & 1 & 7 & 2 & 5 \\ 8 & 7 & 4 & 2 & 4 & 6 & 7 \\ 7 & 9 & 8 & 1 & 9 & 8 & 4 \\ 4 & 7 & 0 & 6 & 4 & 0 & 3 \end{pmatrix}$$

and

$$Q^{-1} = \begin{pmatrix} 1 & 3 & 6 & 7 & 4 & 1 & 0 \\ 1 & 8 & 8 & 3 & 6 & 9 & 8 \\ 6 & 8 & 6 & 8 & 6 & 2 & 1 \\ 3 & 5 & 3 & 5 & 7 & 2 & 2 \\ 5 & 4 & 2 & 3 & 9 & 8 & 6 \\ 1 & 8 & 3 & 7 & 5 & 7 & 2 \\ 6 & 8 & 5 & 5 & 8 & 1 & 4 \end{pmatrix}$$

such that

$$PAQ = \begin{pmatrix} I_4 & 0 \\ 0 & N \end{pmatrix} \quad \text{and} \quad PBQ = \begin{pmatrix} G & 0 \\ 0 & I_3 \end{pmatrix},$$

where

$$N = \begin{pmatrix} 0 & 0 & 0 \\ 1 & 0 & 0 \\ 0 & 1 & 0 \end{pmatrix} \quad \text{and} \quad G = \begin{pmatrix} 3 & 0 & 0 & 0 \\ -1 & 3 & 0 & 0 \\ 0 & -1 & 4 & 0 \\ 0 & 0 & -1 & 4 \end{pmatrix}.$$

Now, the problem is changed into the following IVP:

$$\begin{cases} \dfrac{dy(t)}{dt} = -Gy(t), & t \in (0,1], \\ y(0) = (22, 43, 37, 27)^T. \end{cases}$$

As we did in Example 1, the third order GBDF is used to solve this system of ODEs.

For the GMRES method, only one matrix-vector multiplication with the preconditioned matrix is performed at each iteration. Since T_m in Example 1 is the discrete Laplacian matrix and H_m in Example 2 is the Hamiltonian matrix, there are only few non-zero diagonals in both T_m and H_m. Therefore, the number of operations per iteration is of $O(ms \log s + ms)$ for both cases. Tables 11.1, 11.2 list the number of iterations required for the convergence of the GMRES method for Examples 1, 2 with different m and s. Table 11.3 lists the number of iterations required for the convergence of the GMRES method for Example 3 with different s. For a comparison we also give the number of iterations by the GMRES method with no preconditioner I, the preconditioner $\tilde{c}_F^{(1)}(M)$ given by (2.16) in Section 2.1.4 and Bertaccini's block circulant preconditioner P proposed in [6]. More precisely, the preconditioner P is defined as follows,

$$P = \tilde{A} \otimes I_m - h\tilde{B} \otimes J_m$$

where the diagonals $\tilde{\alpha}_j$ and $\tilde{\beta}_j$ of $\tilde{A}$ and $\tilde{B}$ are given by

$$\tilde{\alpha}_j = \left(1 + \frac{j}{s+1}\right)\alpha_{j+\nu} + \frac{j}{s+1}\alpha_{j+\nu-(s+1)}, \quad j = 0, 1, \cdots, s,$$

and

$$\tilde{\beta}_j = \left(1 + \frac{j}{s+1}\right)\beta_{j+\nu} + \frac{j}{s+1}\beta_{j+\nu-(s+1)}, \quad j = 0, 1, \cdots, s,$$

respectively. We see from Tables 11.1, 11.2, 11.3 that in Examples 1, 2, 3, as s increases, the number of iterations increases if no preconditioner is used. We also find that the performances of the preconditioner $\tilde{s}_F^{(1)}(M)$ are better than those of the preconditioners $\tilde{c}_F^{(1)}(M)$ and P.

Table 11.1. Number of iterations for convergence

m	s	I	$\tilde{s}_F^{(1)}(M)$	$\tilde{c}_F^{(1)}(M)$	P
24	6	9	3	6	6
	12	35	3	7	7
	24	88	3	7	8
	48	176	3	6	8
	96	>200	3	6	8
48	6	17	3	6	6
	12	65	3	7	7
	24	166	3	7	8
	48	>200	3	7	8
	96	>200	3	6	8
96	6	31	3	6	6
	12	129	3	7	7
	24	>200	3	7	8
	48	>200	3	7	8
	96	>200	3	6	8

Example 1

Table 11.2. Number of iterations for convergence

m	s	I	$\tilde{s}_F^{(1)}(M)$	$\tilde{c}_F^{(1)}(M)$	P
24	6	113	43	44	46
	12	149	41	43	44
	24	138	32	34	35
	48	137	23	23	26
	96	167	16	15	20
48	6	217	64	67	70
	12	271	62	65	67
	24	237	55	54	57
	48	226	40	41	44
	96	259	23	21	28
96	6	>1000	96	101	105
	12	>1000	92	96	98
	24	>1000	89	91	95
	48	>1000	69	71	73
	96	>1000	43	41	48

Example 2

Table 11.3. Number of iterations for convergence

s	I	$\tilde{s}_F^{(1)}(M)$	$\tilde{c}_F^{(1)}(M)$	P
6	19	10	12	13
12	26	11	11	13
24	42	10	11	14
48	71	10	12	14
96	130	10	12	14

Example 3

Chapter 12

Applications in Image Processing

In this chapter, we study image restoration problems. The image of an object is modeled as

$$g(\xi, \delta) = s \left\{ \int_{-\infty}^{\infty} \int_{-\infty}^{\infty} t(\xi, \delta; \alpha, \beta) f(\alpha, \beta) d\alpha d\beta \right\} + \eta(\xi, \delta)$$

where $g(\xi, \delta)$ is the degraded image, $f(\alpha, \beta)$ is the original image, $\eta(\xi, \delta)$ represents an additive noise, and $s\{\cdot\}$ represents the non-linear characteristics of the device that senses and records the image. The image restoration problem is that given an observed image g, compute an approximation to the original image f. The regularized PCGLS method with some preconditioners based on fast transforms is applied to solving linear systems arising from image restoration.

12.1   Introduction

Image restoration problems in image processing are the problems of eliminating or minimizing degradations in blurred noisy images. Degradations in a blurred image can be introduced by optical systems, image motion, and noises from electronic and photometric sources. When a recorded image is degraded by blurring and noise, important information remains hidden and cannot be interpreted directly without numerical processing. Such kind of problem occurs in a variety of applications, for instance, acoustical and radar imaging, medical imaging and space activities. Thus, image restoration problems have attracted widespread interest. Here, we only consider the linear model.

The image of an object can be modeled as

$$g(\xi, \delta) = \int_{-\infty}^{\infty} \int_{-\infty}^{\infty} t(\xi, \delta; \alpha, \beta) f(\alpha, \beta) d\alpha d\beta + \eta(\xi, \delta) \qquad (12.1)$$

where $g(\xi, \delta)$ is the recorded (or degraded) image, $f(\alpha, \beta)$ is the ideal (or original) image, $\eta(\xi, \delta)$ represents an additive noise. The function $t(\xi, \delta; \alpha, \beta)$ is called the point spread function (PSF) and represents the degradation of the image. The task of image restoration is: given an observed image g, the function t and, possibly, the statistics of noise η, compute an approximation to the original image f.

In the digital implementation, the integral in (12.1) is discretized by using some quadrature rules to obtain a discrete scalar model

$$g(i, j) = \sum_{k=1}^{n} \sum_{l=1}^{n} t(i, j; k, l) f(k, l) + \eta(i, j).$$

In matrix form the image restoration problem (12.1) can be written as

$$g = Tf + \eta \qquad (12.2)$$

where g, η, $f \in \mathbb{R}^{n^2}$ and $T \in \mathbb{R}^{n^2 \times n^2}$. This is the square image formulation. Two remarks on the linear model (12.2) are given below:

(i) The addition of the non-zero noise vector η, which is given by a random process, may imply that the solution of (12.2) is not unique.

(ii) Usually, T is a large, sparse and ill conditioned matrix and hence is extremely sensitive to noise η.

The remarks indicate that if T has no special structure, then the image restoration problems are extremely complicated to be solved. But in many cases, T has a block Toeplitz structure or a BTTB structure.

For the PSF, usually, $t(\xi, \delta; \alpha, \beta)$ varies with the position in both image and object planes. In this case, the PSF is said to be spatially variant, and the corresponding matrix T in (12.2) will have no special structure. In many practical applications, however, the PSF is spatially invariant, i.e., it can be written as

$$t(\xi, \delta; \alpha, \beta) = t(\xi - \alpha, \delta - \beta).$$

We then have the following model

$$g(\xi, \delta) = \int_{-\infty}^{\infty} \int_{-\infty}^{\infty} t(\xi - \alpha, \delta - \beta) f(\alpha, \beta) d\alpha d\beta + \eta(\xi, \delta). \qquad (12.3)$$

In the digital implementation of (12.3), we have the following discrete scalar model

$$g(i, j) = \sum_{k=1}^{n} \sum_{l=1}^{n} t(i - k, j - l) f(k, l) + \eta(i, j).$$

In matrix form, we have

$$g = Tf + \eta \qquad (12.4)$$

where $T \in \mathbb{R}^{n^2 \times n^2}$ is a BTTB matrix with entries given by

$$t_{i,k;j,l} = t(i - k, j - l),$$

for $i, j, k, l = 1, \cdots, n$ and g, η, $f \in \mathbb{R}^{n^2}$. Since T in (12.4) is very ill conditioned, we need to use techniques of regularization in order to find a reasonably approximate solution.

12.2 Regularized PCGLS method

Tikhonov regularization method is a commonly used method to deal with ill-posed problems, see [22] and [34]. Let X and Y be Hilbert spaces, $T : X \to Y$ be a linear and bounded operator, $x \in X$ and $y \in Y$. Let us consider the following equation:

$$Tx = y.$$

Tikhonov functional is defined as follows, see [34].

Definition 12.1 *For a given $\alpha \in \mathbb{R}$, Tikhonov functional $J_\alpha(x)$ is defined as*

$$J_\alpha(x) \equiv \|Tx - y\|^2 + \alpha \|Lx\|^2 \qquad (12.5)$$

where α is called the regularization parameter and L is the regularization operator.

Then, the standard solution of (12.5) can be obtained by the following theorem, see [34].

Theorem 12.1 *If the regularization operator L is chosen to be the identity operator or some differential operators, and $\alpha > 0$, then Tikhonov functional J_α has a unique minimum $x_\alpha \in X$. Here x_α can be obtained by solving the following normal equation*

$$(\alpha L^* L + T^* T) x = T^* y.$$

In (12.5), the regularization parameter α controls the degree of smoothness of the solution. If α is too large, then the approximate solution is far away from the true solution of $Tx = y$. If α is too small, then the problem will still be quite ill. In general, it is non-trivial to choose the size of α analytically. We then choose a suitable size of α from experiments.

For (12.4), we consider the following LS problem

$$\min ||\tilde{g} - \tilde{T}x(\alpha)||_2. \tag{12.6}$$

Here

$$\tilde{g} = \begin{pmatrix} g \\ 0 \end{pmatrix} \quad \text{and} \quad \tilde{T} = \begin{pmatrix} T \\ \alpha L \end{pmatrix}$$

where L is usually chosen to be the identity matrix I or some difference operator matrices. For instance, the second difference operator L has the following rectangular band-Toeplitz form:

$$L = \begin{pmatrix} 1 & & & & & & \\ -2 & 1 & & & & & \\ 1 & -2 & 1 & & & & \\ & 1 & -2 & 1 & & & \\ & & \ddots & \ddots & \ddots & & \\ & & & & 1 & -2 & 1 \\ & & & & & 1 & -2 \\ & & & & & & 1 \end{pmatrix}.$$

We will use $L = I$ for the examples in Section 12.3.

The PCGLS method is proposed as main tool to solve (12.6). The scheme of the PCGLS method for solving (12.6) is given as follows, see [22]. Let $x^{(0)}$ be an initial approximation to

$$\tilde{T}x = \tilde{g}$$

and let M be a given preconditioner. At the initialization step, we have

$$r^{(0)} = \tilde{g} - \tilde{T}x^{(0)},$$

$$p^{(0)} = s^{(0)} = (M^{-1})^*\tilde{T}^*r^{(0)},$$

$$\gamma_0 = \|s^{(0)}\|_2^2.$$

In each iteration step, we have

$$\begin{cases} q^{(k)} := \tilde{T}M^{-1}p^{(k)}, \\[2mm] \alpha_k := \dfrac{\gamma_k}{\|q^{(k)}\|_2^2}, \\[2mm] x^{(k+1)} := x^{(k)} + \alpha_k M^{-1}p^{(k)}, \\[2mm] r^{(k+1)} := r^{(k)} - \alpha_k q^{(k)}, \\[2mm] s^{(k+1)} := (M^{-1})^*\tilde{T}^*r^{(k+1)}, \\[2mm] \gamma_{k+1} := \|s^{(k+1)}\|_2^2, \\[2mm] \beta_k := \dfrac{\gamma_{k+1}}{\gamma_k}, \\[2mm] p^{(k+1)} := s^{(k+1)} + \beta_k p^{(k)}, \end{cases}$$

where $q^{(k)}$, $p^{(k)}$, $r^{(k)}$ and $s^{(k)}$ are vectors, and α_k, β_k and γ_k are scalars, for $k = 0, 1, 2, \cdots$. The $x^{(k)}$ is the approximation to the true solution after the k-th iteration.

12.3 Numerical results

Restoration of real image by using the regularized PCGLS algorithm has been carried out. The datum we used are generated from two source images, see Fig. 12.1 (a) and Fig. 12.5 (a).

Example 1. The first test example is a 128×128 image of a woman face, see Fig. 12.1 (a). We consider the spatially invariant discretized matrix T with entries given by

$$t_{i-k,j-l} = \begin{cases} \exp\{-800(i-k)^2-800(j-l)^2\}, & \text{if } \sqrt{(i-k)^2+(j-l)^2} < 1/8, \\ 0, & \text{otherwise.} \end{cases}$$

(12.7)

The size of T is 16384×16384. The observed image is constructed by forming the vector

$$g = Tf + \eta$$

where T is defined by (12.7) and f is a vector formed by row ordering the original image. By unstacking the vector g, we obtain the blurred noisy image, see Fig. 12.1 (b). The noise function η is from the normal distribution and is scaled such that $\|\eta\|_2/\|Tf\|_2 = 10^{-4}$.

Our goal is to recover an approximation to the original image f when g and T are given. Since T is very ill conditioned, we use the regularized PCGLS method proposed in Section 12.2 with preconditioners S and C, where the preconditioners S and C are defined to be the minimizers of $\|T - B\|_F$ over all $B \in \mathcal{M}_{\Phi_n^s \otimes \Phi_n^s}$ and $\mathcal{M}_{\Phi_n^c \otimes \Phi_n^c}$, respectively, see Section 7.2. The convergence results for Example 1 with no preconditioner, the preconditioners S and C are given in Figs. 12.2, 12.3 and 12.4, respectively.

Example 2. The second test example is a 256×256 image of an ocean reconnaissance satellite, see Fig. 12.5 (a). We then consider the spatially invariant discretized matrix T with entries given by

$$t_{i-k,j-l} = \begin{cases} \exp\{-800(i-k)^2-800(j-l)^2\}, & \text{if } |i-k|, |j-l| < 1/8, \\ 0, & \text{otherwise.} \end{cases}$$

The size of T is 65536×65536. We then have the following observed image

$$g = Tf + \eta,$$

see Fig. 12.5 (b).

The convergence results of the regularized PCGLS method for Example 2 with no preconditioner, the preconditioners S and C are given in Figs. 12.6, 12.7 and 12.8, respectively.

From these test examples, we note that the restorations with the pre-conditioner C are fairly good after a few iterations, see Figs. 12.4 and 12.8. In contrast, the restorations with no preconditioner are not good enough until several more iterations are completed, see Figs. 12.2 and 12.6.

Fig. 12.1   (a) True image,   (b) Observed image.

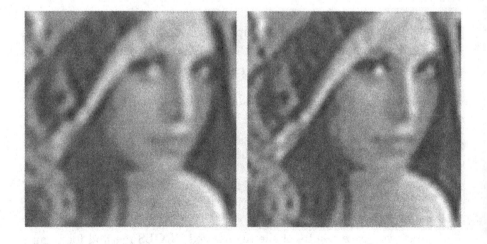

Fig. 12.2   (a) 30 iterations,   **(b)** 174 **iterations.**

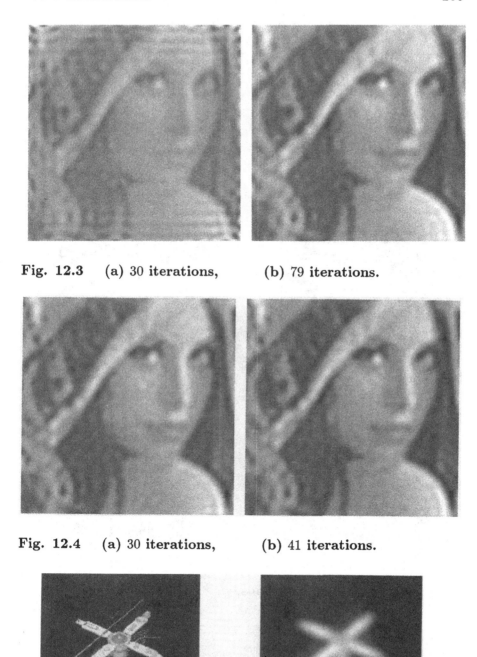

Fig. 12.3   (a) 30 iterations,   (b) 79 iterations.

Fig. 12.4   (a) 30 iterations,   (b) 41 iterations.

Fig. 12.5   (a) Original image,   (b) Observed image.

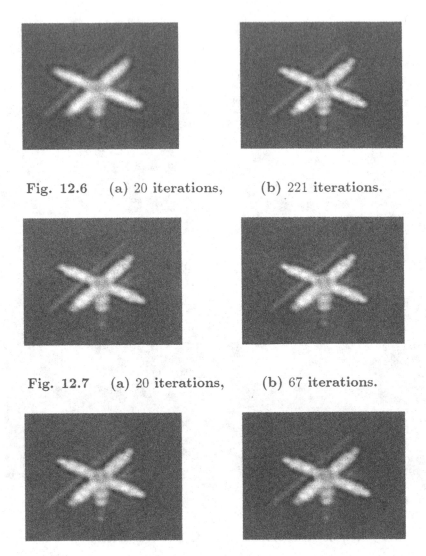

Fig. 12.6   (a) 20 iterations,   (b) 221 iterations.

Fig. 12.7   (a) 20 iterations,   (b) 67 iterations.

Fig. 12.8   (a) 10 iterations,   (b) 22 iterations.

Bibliography

[1] G. Ammar and W. Gragg, *Superfast Solution of Real Positive Definite Toeplitz Systems*, SIAM J. Matrix Anal. Appl., Vol. 9 (1988), pp. 61-76.

[2] O. Axelsson, *Iterative Solution Methods*, Cambridge University Press, Cambridge (1996).

[3] F. Benedetto, *Analysis of Preconditioning Techniques for Ill Conditioned Toeplitz Matrices*, SIAM J. Sci. Comput., Vol. 16 (1995), pp. 682-697.

[4] F. Benedetto, *Preconditioning of Block Toeplitz Matrices by Sine Transforms*, SIAM J. Sci. Comput., Vol. 18 (1997), pp. 499-515.

[5] F. Benedetto and S. Serra, *A Unifying Approach to Matrix Algebra Preconditioning*, Numer. Math., Vol. 82 (1999), pp. 57-90.

[6] D. Bertaccini, *A Circulant Preconditioner for the Systems of LMF-based ODE Codes*, SIAM J. Sci. Comput., Vol. 22 (2000), pp. 767-786.

[7] D. Bini and F. Benedetto, *A New Preconditioner for the Parallel Solution of Positive Definite Toeplitz Systems*, Proceedings of the 2nd AMC Symposium on Parallel Algorithms and Architectures (1990), pp. 220-223.

[8] D. Bini and P. Favati, *On a Matrix Algebra Related to the Discrete Hartley Transform*, SIAM J. Matrix Anal. Appl., Vol. 14 (1993), pp. 500-507.

[9] D. Bini and B. Meini, *Effective Methods for Solving Banded Toeplitz Systems*, SIAM J. Matrix Anal. Appl., Vol. 20 (1999), pp. 700-719.

[10] A. Böttcher and B. Silbermann, *Introduction to Large Truncated Toeplitz Matrices*, Springer, New York (1999).

[11] K. Brenan, S. Campbell and L. Petzold, *Numerical Solution of Initial-Value Problems in Differential-Algebraic Equations*, SIAM Press, Philadelphia (1996).

[12] W. Briggs, V. Henson and S. McCormick, *A Multigrid Tutorial*, 2nd edition, SIAM Press, Philadelphia (2000).

[13] L. Brugnano and D. Trigiante, *Solving Differential Problems by Multistep Initial and Boundary Value Methods*, Gordon and Berach Science Publishers, Amsterdam (1998).

[14] R. Chan, *Circulant Preconditioners for Hermitian Toeplitz Systems*, SIAM J. Matrix Anal. Appl., Vol. 10 (1989), pp. 542-550.

[15] R. Chan, *Numercal Solutions for the Inverse Heat Problems in R^N*, The SEAMS Bull. Math., Vol. 16 (1992), pp. 97-105.

[16] R. Chan, *Toeplitz Preconditioners for Toeplitz Systems with Non-negative Generating Functions*, IMA J. Numer. Anal., Vol. 11 (1991), pp. 333-345.

[17] R. Chan and T. Chan, *Circulant Preconditioners for Elliptic Problems*, Numer. Linear Algebra Appl., Vol. 1 (1992), pp. 77-101.

[18] R. Chan, T. Chan and C. Wong, *Cosine Transform Based Preconditioners for Total Variation Deblurring*, IEEE Trans. Image Proc., Vol. 8 (1999), pp. 1472-1478.

[19] R. Chan, Q. Chang and H. Sun, *Multigrid Method for Ill Conditioned Symmetric Toeplitz Systems*, SIAM J. Sci. Comput., Vol. 19 (1998), pp. 516-529.

[20] R. Chan and X. Jin, *A Family of Block Preconditioners for Block Systems*, SIAM J. Sci. Statist. Comput., Vol. 13 (1992), pp. 1218-1235.

[21] R. Chan, X. Jin and M. Yeung, *The Circulant Operator in the Banach Algebra of Matrices*, Linear Algebra Appl., Vol. 149 (1991), pp. 41-53.

[22] R. Chan, J. Nagy and R. Plemmons, *FFT-Based Preconditioners for Toeplitz-Block Least Squares Problems*, SIAM J. Numer. Anal., Vol. 30 (1993), pp. 1740-1768.

[23] R. Chan and M. Ng, *Conjugate Gradient Methods for Toeplitz Systems*, SIAM Review, Vol. 38 (1996), 427-482.

[24] R. Chan, M. Ng and X. Jin, *Strang-Type Preconditioners for Systems of LMF-Based ODE Codes*, IMA J. Numer. Anal., Vol. 21 (2001), pp. 451-462.

[25] R. Chan, M. Ng and C. Wong, *Sine Transform Based Preconditioners for Symmetric Toeplitz Systems*, Linear Algebra Appl., Vol. 232 (1996), pp. 237-259.

[26] R. Chan and G. Strang, *Toeplitz Equations by Conjugate Gradients with Circulant Preconditioner*, SIAM J. Sci. Statist. Comput., Vol. 10 (1989), pp. 104-119.

[27] R. Chan and P. Tang, *Fast Band-Toeplitz Preconditioner for Hermitian Toeplitz Systems*, SIAM J. Sci. Comput., Vol. 15 (1994), pp. 164-171.

[28] R. Chan and M. Yeung, *Circulant Preconditioners for Toeplitz Matrices with Positive Continuous Generating Functions*, Math. Comput., Vol. 58 (1992), pp. 233-240.

[29] R. Chan and M. Yeung, *Circulant Preconditioners Constructed from Kernels*, SIAM J. Numer. Anal., Vol. 29 (1992), pp. 1093-1103.

[30] T. Chan, *An Optimal Circulant Preconditioner for Toeplitz Systems*, SIAM J. Sci. Statist. Comput., Vol. 9 (1988), pp. 766-771.

[31] T. Chan and J. Olkin, *Preconditioners for Toeplitz-Block Matrices*, Numer. Algorithms, Vol. 6 (1994), pp. 89-101.

[32] Q. Chang, Y. Wong and H. Fu, *On the Algebraic Multigrid Method*, J. Comput. Physics, Vol. 125 (1996), pp. 279-292.

[33] P. Davis, *Circulant Matrices*, John Wiley & Sons, Inc., New York (1979).

[34] H. Engl, M. Hanke and A. Neubauer, *Regularization of Inverse Problems*, Kluwer Academic Publishers, Netherlands (1996).

[35] G. Fiorentino and S. Serra, *Multigrid Methods for Toeplitz Matrices*, Calcolo, Vol. 28 (1991), pp. 283-305.

[36] G. Fiorentino and S. Serra, *Multigrid Methods for Symmetric Positive Definite Block Toeplitz Matrices with Non-negative Generating Functions*, SIAM J. Sci. Comput., Vol. 17 (1996), pp. 1068-1081.

[37] D. Gilliam, C. Martin and J. Lund, *Analytic and Numerical Aspects of the Observation of the Heat Equation*, Proceedings of the 26th IEEE Conference on Decision and Control (1987), pp. 975-976.

[38] G. Golub and C. van Loan, *Matrix Computations*, 3rd edition, Johns Hopkins Univ. Press, Baltimore (1996).

[39] R. Gonzalez and R. Woods, *Digital Image Processing*, Addison-Wesley Publishing Company, Inc., New York (1992).

[40] U. Grenander and G. Szegö, *Toeplitz Forms and Their Applications*, 2nd edition, Chelsea, New York (1984).

[41] M. Gulliksson, *Iterative Refinement for Constrained and Weighted Linear Least Squares*, BIT, Vol. 34 (1994), pp. 239–253.

[42] M. Gulliksson, X. Jin and Y. Wei, *Perturbation Bounds for Constrained and Weighted Linear Least Squares Problems*, Linear Algebra Appl., Vol. 349 (2002), pp. 221–232.

[43] G. Heinig and K. Rost, *Algebraic Methods for Toeplitz-Like Matrices and Operators*, Birkhäuser, Boston (1984).

[44] L. Hemmingsson and K. Otto, *Analysis of Semi-Toeplitz Preconditioners for First Order PDE*, SIAM J. Sci. Comput., Vol. 17 (1996), pp. 47-64.

[45] S. Holmgren and K. Otto, *Iterative Solution Methods and Preconditioners for Block Tri-diagonal Systems of Equations*, SIAM J. Matrix Anal. Appl., Vol. 13 (1992), pp. 863-886.

[46] S. Holmgren and K. Otto, *Semi-Circulant Preconditioners for First Order Partial Differential Equations*, SIAM J. Sci. Comput., Vol. 15 (1994), pp. 385-407.

[47] S. Holmgren and K. Otto, *Analysis of Preconditioners for Hyperbolic Partial Differential Equations*, SIAM J. Numer. Anal., Vol. 33 (1996), pp. 2131-2165.

[48] S. Holmgren and K. Otto, *A Framework for Polynomial Preconditioners Based on Fast Transform I: Theory*, BIT, Vol. 38 (1998), pp. 544-559.

[49] S. Holmgren and K. Otto, *A Framework for Polynomial Preconditioners Based on Fast Transform II: PDE Applications*, BIT, Vol. 38 (1998), pp. 721-736.

[50] T. Huckle, *Circulant and Skew Circulant Matrices for Solving Toeplitz Matrix Problems*, SIAM J. Matrix Anal. Appl., Vol. 13 (1992), pp. 767-777.

[51] X. Jin, *Sine Transform Preconditioners for Second Order Partial Differential Equations*, Numer. Math. J. Chinese Univ., Vol. 2 (1993), pp. 116-123.

[52] X. Jin, *Hartley Preconditioners for Toeplitz Systems Generated by Positive Continuous Functions*, BIT, Vol. 34 (1994), pp. 367-371.

[53] X. Jin, *A Note on Preconditioned Block Toeplitz Matrices*, SIAM J. Sci. Comput., Vol. 16 (1995), pp. 951-955.

[54] X. Jin, *A Fast Algorithm for Block Toeplitz Systems with Tensor Structure*, Appl. Math. Comput., Vol. 73 (1995), pp. 115-124.

[55] X. Jin, *Fast Iterative Solvers for Symmetric Toeplitz Systems – A Survey and an Extension*, J. Comput. Appl. Math., Vol. 66 (1996), pp. 315-321.

[56] X. Jin, *A Preconditioner for Constrained and Weighted Least Squares Problem with Toeplitz Structure*, BIT, Vol. 36 (1996), pp. 101-109.

[57] X. Jin, *Band-Toeplitz Preconditioners for Block Toeplitz Systems*, J. Comput. Appl. Math., Vol. 70 (1996), pp. 225-230.

[58] X. Jin, *A Note on Construction of Circulant Preconditioners from Kernels*, Appl. Math. Comput., Vol. 83 (1997), pp. 3-12.

[59] X. Jin and R. Chan, *Circulant Preconditioners for Second Order Hyperbolic Equations*, BIT, Vol. 32 (1992), pp. 650-664.

[60] X. Jin and S. Lei, *Sine Transform Based Preconditioners for Solving Constant-Coefficient First-Order PDEs*, Linear Algebra Appl., to appear.

[61] F. John, *Partial Differential Equations*, 4th edition, Springer, New York (1982).

[62] T. Kailath and A. Sayed, eds., *Fast Reliable Algorithms for Matrices with Structures*, SIAM Press, Philadelphia (1999).

[63] H. Keller, *A New Difference Scheme for Parabolic Problems*, Numerical Solution of Partial Differential Equations-II, pp. 327-350, Academic Press, New York (1971).

[64] K. Kou, V. Sin and X. Jin, *A Note on the Fast Algorithm for Block Toeplitz Systems with Tensor Structure*, Appl. Math. Comput., Vol. 126 (2002), pp. 187-197.

[65] T. Ku and C. Kuo, *On the Spectrum of a Family of Preconditioned Block Toeplitz Matrices*, SIAM J. Sci. Statist. Comput., Vol. 13 (1992), pp. 948-966 .

[66] L. Lapidus and G. Pinder, *Numerical Solution of Partial Differential Equations in Science and Engineering*, Wiley, New York (1982).

[67] S. Lei and X. Jin, *Strang-Type Preconditioners for Solving Differential-Algebraic Equations*, NAA 2000, LNCS, Vol. 1988, pp. 505-512, eds., L. Vulkov, J. Wasniewski and P. Yalamov, Springer, Berlin (2001).

[68] S. Lei, K. Kou and X. Jin, *Preconditioners for Ill Conditioned Block Toeplitz Systems with Application in Image Restoration*, East-West J. Numer. Math., Vol. 7 (1999), pp. 175-185.

[69] M. Ng, *Band Preconditioners for Block-Toeplitz-Toeplitz-Block-Systems*, Linear Algebra Appl., Vol. 259 (1997), pp. 307-327.

[70] J. Olkin, *Linear and Non-linear Deconvolution Problems*, Ph.D. thesis, Rice University, Houston, Texas (1986).

[71] D. Potts and G. Steidl, *Preconditioners for Ill Conditioned Toeplitz Matrices*, BIT, Vol. 39 (1999), pp. 579-594.

[72] W. Rudin, *Functional Analysis*, McGraw-Hill, Inc, New York (1991).

[73] W. Rudin, *Real and Complex Analysis*, McGraw-Hill, Inc, New York (1986).

[74] J. Ruge and K. Stuben, *Algebraic Multigrid*, Multigrid Methods, pp. 73-130, ed., S. McCormick, SIAM Press, Philadelphia (1987).

[75] Y. Saad, *Iterative Methods for Sparse Linear Systems*, PWS Publishing Company, Boston (1996).

[76] M. Seager, *Parallelizing Conjugate Gradient for the CRAY X-MP*, Parallel Comuting, Vol. 3 (1986), pp. 35-47.

[77] S. Serra, *Preconditioning Strategies for Asymptotically Ill Conditioned Block Toeplitz Systems*, BIT, Vol. 34 (1994), pp. 579-594.

[78] S. Serra, *Optimal, Quasi-Optimal and Superlinear Band-Toeplitz Preconditioners for Asymptotically Ill Conditioned Positive Definite Toeplitz Systems*, Math. Comp., Vol. 65 (1997), pp. 651-665.

[79] S. Serra, *On the Extreme Eigenvalues of Hermitian (Block) Toeplitz Matrices*, Linear Algebra Appl., Vol. 270 (1998), pp. 109-129.

[80] S. Serra, *Toeplitz Preconditioners Constructed from Linear Approximation Processes*, SIAM J. Matrix Anal. Appl., Vol. 20 (1998), pp. 446-465.

[81] S. Serra, *How to Choose the Best Iterative Strategy for Symmetric Toeplitz Systems*, SIAM J. Numer. Anal., Vol. 36 (1999), pp. 1078-1103.

[82] S. Serra and C. Tablino, *Spectral and Structural Analysis of High Order Finite Difference Matrices Discretizing Elliptic Operators*, Linear Algebra Appl., Vol. 293 (1999), pp. 85-131.

[83] S. Serra and E. Tyrtyshnikov, *Any Circulant-Like Preconditioner for Multilevel Matrices Is Not Superlinear*, SIAM J. Matrix Anal. Appl., Vol. 21 (1999), pp. 431-439.

[84] M. Shirvani and J. So, *Solutions of Linear Differential Algebraic Equations*, SIAM Review, Vol. 40 (1998), 344-346.

[85] P. Sonneveld, *GGS, a Fast Lanczos-type Solver for Non-symmetric Linear Systems*, SIAM J. Sci. Statist. Comput., Vol. 10 (1989), pp. 36-52.

[86] F. Stenger, *Numerical Methods Based on the Whittaker Cardinal, or Sinc Functions*, SIAM Review, Vol. 23 (1981), pp. 165-224.

[87] G. Strang, *A Proposal for Toeplitz Matrix Calculations*, Stud. Appl. Math., Vol. 74 (1986), pp. 171-176.

[88] H. Sun, R. Chan and Q. Chang, *A Note on the Convergence of the Two-Grid Method for Toeplitz Systems*, Computers Math. Appl., Vol. 34 (1997), pp. 11-18.

[89] H. Sun, X. Jin and Q. Chang, *Convergence of the Multigrid Method for Ill Conditioned Block Toeplitz Systems*, BIT, Vol. 41 (2001), pp. 179-190.

[90] P. Swarztrauber, *Multiprocessor FFTs*, Parallel Comuting, Vol. 5 (1987), pp. 197-210.

[91] G. Szegö, *Orthogonal Polynomials*, AMS Colloquium Publications (1939).

[92] R. Thompson, *Principal Submatrices IX: Interlacing Inequalities for Singular Values of Submatrices*, Linear Algebra Appl., Vol. 5 (1972), pp. 1-12.

[93] P. Tilli, *Singular Values and Eigenvalues of Non Hermitian Block Toeplitz Matrices*, Linear Algebra Appl., Vol. 272 (1998), pp. 59-89.

[94] P. Tilli, *Locally Toeplitz Matrices: Spectral Theory and Applications*, Linear Algebra Appl., Vol. 278 (1998), pp. 91-120.

[95] L. Trefethen and D. Bau, *Numerical Linear Algebra*, SIAM Press, Philadelphia (1997).

[96] E. Tyrtyshnikov, *Optimal and Super-Optimal Circulant Preconditioners*, SIAM J. Matrix Anal. Appl., Vol. 13 (1992), pp. 459-473.

[97] E. Tyrtyshnikov, *Circulant Preconditioners with Unbounded Inverses*, Linear Algebra Appl., Vol. 216 (1995), pp. 1-23.

[98] E. Tyrtyshnikov, *A Unifying Approach to Some Old and New Theorems on Distribution and Clustering*, Linear Algebra Appl., Vol. 232 (1996), pp. 1-43.

[99] H. van der Vorst, *Preconditioning by Incomplete Decompositions*, Ph.D. thesis, Rijksuniversiteit te Utrecht, Utrecht (1982).

[100] J. Wilkinson, *The Algebraic Eigenvalue Problem*, Clarendon Press, Oxford (1965).

[101] A. Zygmund, *Trigonometric Series*, Vol. I, Cambridge University Press, Cambridge (1968).

Index